TECHNOLOGY LEGISLATION
IS COMING.

Technology Legislation is Coming.

Is Your Business Ready?

Chris Wiser

Published by Game Changer Publishing

Paperback ISBN: 978-1-965653-74-6

Hardcover ISBN: 978-1-965653-75-3

Digital ISBN: 978-1-965653-77-7

GAME CHANGER PUBLISHING

www.GameChangerPublishing.com

TECHNOLOGY LEGISLATION IS COMING.

IS YOUR BUSINESS READY?

FOREWORD

REGULATION IS COMING

In today's digital age, cybersecurity is no longer an option but a necessity. Cybercriminals have been quietly capitalizing on small and medium-sized businesses for years. These businesses are low-hanging fruit for hackers due to their general lack of proper protection and the fact that many of these business owners don't see themselves as targets. This is largely because businesses of this size rarely make the news or even discuss things like this with other business owners and, therefore, do not believe they are impacted.

Unfortunately, this has contributed to attacks increasing, cybersecurity insurance rates rising, while protections decrease, and exclusions continue to get added. Cybersecurity Insurance questionnaires have gone from nothing, to a page or two, to 12+ pages from some carriers. New regulations related to cybersecurity have come down from the SEC and FTC in the United States, impacting many businesses. With every new cyberattack, we inch closer and closer to regulation on cybersecurity across the board, and many businesses are not prepared for it.

This book is a timely and essential guide for navigating the complex world of cybersecurity regulation. Written by a team of cybersecurity experts, each chapter delves into specific aspects of the subject, providing you with practical insights and actionable advice. Our goal is to equip you with the knowledge and tools you need to safeguard your business, comply with emerging regulations, and thrive in a secure digital environment.

As you embark on this journey, remember that cybersecurity is not just a technical issue; it is a business imperative. By understanding and implementing the guidance within these pages, you will not only protect your assets but also build trust with your customers, partners, and stakeholders.

We hope this book serves as a valuable resource, empowering you to make informed decisions and stay ahead of the curve. Cybersecurity regulation is coming, and with the right preparation, you can turn this challenge into an opportunity for growth and resilience.

Chris Wiser

CEO, 7 Figure MSP™

Speaker/Trainer/Entrepreneur Coach

CONTENTS

ONE
THINGS HAVE CHANGED
THE EVOLUTION OF CYBER THREATS

Author: Nick Stevens
Company: Heroic Technologies

THINGS HAVE CHANGED... wow, what an understatement. In the early 2000s, whether you were working in an organization or an individual, you didn't have much to worry about regarding cyber threats. Most weren't threats at all; they were more of a nuisance. It would be common to see spyware toolbars in Microsoft's Internet Explorer or other adware that caused obnoxious popups. Thankfully, the actual risk from these toolbars and unwanted adware programs was low, and the tools to remove them were readily available and generally free. This made cybersecurity seem very one-dimensional and, thus, a very manageable challenge. If organizations could control permissions for users, applications, and data, they would be able to keep computers free from spyware, and the likelihood of cybersecurity events would remain relatively low.

Fast forward to today, and the threats seem to be everywhere and look much different. Not only have they evolved in complexity, but the frequency of attacks has also increased, and the impact on organizations and individuals has increased exponentially. Now, businesses must make an intentional effort to consider cyber risk as a part of all of their business decisions and overall business strategy, as cybersecurity has become a central part of a business's operations.

Together, we'll walk through the evolution of cyber threats as well as some of the most significant cyberattacks over the past two decades. We'll explore how cybercriminals have shifted their tactics as a response to organizations and security professionals working to improve their defenses. We'll also discuss the critical need for organizations to continue focusing on risk management and ensuring defenses are continuously improved, lest they risk potentially devasting cyberattacks.

THE DAWN OF A NEW THREAT LANDSCAPE FOR BUSINESSES: EARLY 2000S TO 2012

As we previously covered, the threats of the early 2000s were often relatively unsophisticated compared to those we're seeing today. Common threats included spyware toolbars, which might frustrate employees and everyday computer users, but they rarely caused significant issues to a business. We also started to see the initial creation of some of the first botnets or large groups of compromised computers; these computers would then be used for Distributed Denial of Service (DDoS) attacks. An attack occurs when an extremely large volume of information or requests is sent to a specific resource like a website or router, so that resource becomes overwhelmed and unresponsive, essentially taking it offline. This would allow a cybercriminal to target a business's public website or firewall and potentially take it offline, temporarily disrupting its operations. While these can be very damaging, these attacks were few and far between in the 2000s. However, as we moved toward the end of the decade, we saw a far more dangerous threat emerge: Advanced Persistent Threats (APTs).

Advanced Persistent Threats emerged as one of the most notable cybersecurity events in this period. APTs are very well-funded and organized groups, often associated with nation-states or organized crime. They carry out carefully planned attacks that can span a long period, engaging a variety of tactics that allow them to be stealthy and persistent within an organization's computer network. The goal of an APT may be to disrupt, but it may also be to steal sensitive information, like intellectual property. And the more persistent the threat, the better they would be at stealing information, as they'd often be able to remain in a business's network for an extended period of time without being detected.

This was the case when a highly sophisticated attack, which came to be named Operation Aurora, targeted Google in 2009. While previous cyberattacks often took a seemingly random approach by targeting anyone and everyone, this campaign was different as the victims were specifically targeted. The group behind the campaign was also different, as Operation Aurora was conducted by well-funded, China-based hackers. Attackers had a very specific plan in mind, which included not only accessing Google's software source code but also making changes to it. This was a serious shift in how businesses were targeted, and businesses were immediately aware of a huge increase in risk to their operations, forcing them to invest more resources into cybersecurity, as it was clear threats were going to remain serious and persistent.

RANSOMWARE AND BUSINESS EMAIL COMPROMISE: 2013-2016

In the period from 2013 to 2016, we saw ransomware and business email compromise (BEC) attacks come onto the scene as some of the more threatening risks for businesses. These methods demonstrated an escalation in the techniques cybercriminals used on organizations, including the average small business. There was also a significant escalation in the impact these attacks had on small businesses compared to attacks small businesses previously experienced. Ransomware, as you may know, is a specific malware that encrypts a victim's data and files and then presents a ransom note. If a business wants access to its files and to be able to use its computer systems again, they have one of two options: it can restore from backups if it has complete backups, or it can pay the ransom and hope the hackers will decrypt the files, and they'll be able to work again.

In 2013, Target was breached after its HVAC vendor was hacked via emails loaded with malware. The hackers first compromised the HVAC vendor and then stole credentials and gained access to a "bill, contract submission, and project management" platform used by Target. From this system, they were able to gain further access to Target's network, leading to the theft of 40 million credit card records. In total, it's estimated that this breach will cost Target and insurers over $200 million. This attack was novel at the time because the malicious emails for the initial compromise were directed at one of Target's vendors, making this a supply chain attack, a cyber-attack method that wouldn't become popular for some time later.

During this period, we saw criminals continue to fine-tune their social engineering skills, making business email compromise a significant threat. It became increasingly common for employees to be tricked into transferring funds or revealing confidential or otherwise sensitive information through seemingly legitimate email requests. The financial damage and the reputational impact of these attacks were huge, especially considering that small to medium businesses rarely used cyber-liability insurance during this period. As a result, organizations had to rely on employee training and vigilance, as well as email security, to reduce their risk.

THE ERA OF INTERNET OF THINGS (IoT) AND EARLY AI THREATS: 2017-2020

The next notable shift came between 2017 and 2020 as we started to see the Internet of Things (IoT) and the first artificial intelligence (AI) threats. Cybercriminals saw the explosion of connected IoT devices, whether those are consumer devices like a doorbell camera or internet-connected sensors, thermostats, and other equipment for business, manufacturing, agriculture, healthcare, and so on. While connecting these devices and equipment to the network and the internet made these devices easier to manage and monitor for a business, it also opened a new area of risk that previously didn't exist, as these devices often have little to no security built into them.

In 2017, the infamous WannaCry ransomware attack was launched by North Korean hackers and was arguably the largest of all ransomware attacks at the time this book was published (2024). WannaCry infected 200,000 computers, with over 7,000 infected in the first hour alone. By the time the attack subsided, it had impacted computers in 150 countries and crippled a variety of organizations, from hospitals to corporations, causing an estimated $4 billion in total damages.

On the heels of WannaCry was the NotPetya attack, another highly destructive ransomware strain. The fascinating part of NotPetya is it was believed to be targeted against Ukraine, likely by

Russian hackers. It was also believed to have used Ukrainian accounting software called MeDoc after hackers were able to hijack the software update servers for MeDoc and create a false patch that would deploy the NotPetya ransomware. When the ransomware was deployed, it quickly spread beyond Ukraine, as many of the impacted companies had VPN connections between their offices, including offices in Ukraine. The malware spread across these VPN connections and was able to impact major firms like Maersk and Merck, as well as banks, airports, and power companies.

The Equifax breach, which exposed the information of roughly 150 million people, was another significant incident during this period. This breach illustrated how even large organizations can be overly lax in their security practices, as it was found that Equifax had failed to patch a widely known vulnerability in a consumer complaint portal. Furthermore, this portal wasn't properly segmented from other parts of Equifax's network. Attackers were able to exploit the fact that Equifax failed to renew an encryption certificate for a critical security tool, which allowed the exfiltration of huge volumes of data. The cherry on top was Equifax learned of the breach and delayed notification.

During this period, AI started to play a role in the world of cybersecurity. This gave new tools the ability to better detect and respond to threats. However, it also gave cybercriminals greater abilities to launch even more sophisticated attacks. AI could be used for phishing campaigns and malware, allowing these threats to adapt and change in real-time, which gave them the ability to evade legacy antivirus and anti-spam filters, which used definitions that essentially look for patterns in files. This only further increased the need for organizations to continuously adapt their risk mitigation strategies to ensure they effectively protected themselves.

THE RISE OF RANSOMWARE-AS-A-SERVICE AND SUPPLY CHAIN ATTACKS: 2021-2022

In 2021, the business and cybersecurity world was introduced to Ransomware-as-a-Service and consistent supply chain attacks. Those on the East Coast of the United States likely remember the Colonial Pipeline ransomware attack, which happened in 2021, as this attack brought ransomware to the front page for almost everyone in the U.S. However, those on the East Coast were particularly impacted, as the attack disrupted supplies for fuel, spiking gas prices, delaying shipping and deliveries, and ultimately causing shortages of other goods for American families and businesses. The Colonial Pipeline attack also made it clear that critical infrastructure, including energy, water, healthcare, and so on, is vulnerable to attack, but cyberattacks on this infrastructure have massive consequences that cannot be ignored.

Ransomware-as-a-Service or RaaS are platforms that democratize ransomware or make it available to the masses—cybercriminals—simply by joining an affiliate program. These services hosted on the dark web allow even unskilled hackers to execute cyberattacks that they'd otherwise never be able to accomplish. They can now purchase a "done for you" ransomware campaign through various payment methods, including one-time purchases, monthly subscriptions, affiliate models, and profit sharing. As a part of the campaign, they get access to malware, technical services, support, communication channels, infrastructure for carrying out attacks, and potentially even support with ransom negotiations. This model made it so instead of a small group of elite hackers targeting businesses, there are now huge numbers of attackers, which greatly increased the volume and variety of ransomware incidents.

Supply chain attacks have also gained prominence. In 2021, IT management software provider Kaseya was attacked, which showed how effective a supply chain attack can be when executed properly. Kaseya provides software for corporate IT departments and managed IT service providers to manage computers and computer networks at scale. When hackers were able to compromise Kaseya, they were able to use the software to push out ransomware to hundreds of organizations that used Kaseya's software, shutting down organizations on five different continents, including public schools in New Zealand and a grocery chain in Sweden, on top of hundreds of businesses in the U.S.

THE AGE OF AI: 2022-2024 (AND BEYOND)

As we go from 2022 to 2024, the year this book was published, we start to see the rise of yet another significant threat, perhaps the most significant to date: AI or Artificial Intelligence. Technically, AI was in use before, but the level of threats and risks it brought with it was never at this level. The introduction of publically available Generative AI saw an explosion in use like no technology before, reaching 100 million users in 2-months. It also saw users uploading sensitive information to the platform, including company-owned Intellectual Property, causing some organizations to block ChatGPT and attempt to block other GenAI tools.

When organizations introduced AI tools internally but didn't have proper data governance practices, they quickly found out the tools hurt more than they helped. First, garbage data led to garbage results from the AI systems. Second, if data and files were secured with proper permissions or access control lists (ACLs), employees would be able to ask the AI questions, and it would look up the answer on all the data it could access, and they could access. This resulted in issues like being able to ask what other employees, including management or the CEO, get paid within the organization. This could just as easily be the performance report, home address, or other sensitive information for other employees.

Next, it's fake versions of people, known as deepfakes, which are a version of computer-generated identity fraud. Deepfakes are created using AI software to create realistic-looking but fake videos or photos, as well as audio that can imitate a real person. These can be used to trick people into divulging information or transferring money to fraudulent accounts, as they think the request is coming from a request that they've visually or audibly verified to be legitimate.

Additionally, AI models themselves can be targeted, with a variety of attacks and a variety of reasons—model poisoning, data theft, and prompt injections, to name a few.

In 2023, the DNA testing and ancestry firm 23andMe was breached. Hackers were able to steal personally identifiable information (PII) and genetic profiles. This is, yes, another first, as now someone's genetic information is potentially held by another person, which raises a lot of security concerns and privacy questions on what they can actually do with that type of information.

2023 closed out with one final huge breach when MGM Resorts was taken down by a phone call. That's right; social engineering was supposedly used against MGM's IT helpdesk to get access to employees' credentials, which gave hackers access to MGM's systems. The hacker groups ALPHV and Scattered Spider are reported to have then launched a massive ransomware attack against MGM Resort's systems, forcing MGM to take the systems offline throughout the world, causing patrons to be locked out of their rooms and casinos to have to close. The attack is estimated to have cost MGM over $100 million.

THE COST OF CYBERCRIME: 2023 AND BEYOND

It's easy to see that cybercrime's impact has exploded; in 2023, it was estimated that the total global cost of cybercrime was $11.5 trillion, and that's expected to rise to $14.57 by 2024. At $11.5 trillion, cybercrime would be the third largest economy in the world, only smaller than the United States and China but growing at a rate of almost 27%. The financial impact is staggering, but that's only one part of the total story; you can't ignore the disruptions to organizational operations, reputational damage, or the loss of customer trust, as these are massively impactful to businesses.

In 2024, one of the costliest incidents to date occurred: the ransomware attack on Change Healthcare (a UnitedHealth Group subsidiary). The cyberattack impacted thousands of hospitals, clinics, and pharmacies, stopping them from processing insurance payments. Additionally, 131 million patients may have had their records impacted as a result of the compromise. The attack was made possible because basic and, frankly, fundamental cybersecurity measures like multi-factor authentication (MFA) were not enabled on remote access systems.

As this book is being written, another major ransomware attack is unfolding against CDK Global, a company that provides software that is critical to the operations of over 15,000 automotive dealerships in North America. As CDK attempted to restore its systems, a second ransomware attack was launched, shutting them down once again. CDK is now considering paying a substantial ransom to restore its systems. Otherwise, its clients will have to continue running their operations with pen and paper.

THE MULTIFACETED THREATS OF TODAY

Today's cyber threats are no longer confined to a single vector like they used to be when an attack might take place simply over email. Modern attacks leverage a variety of techniques, from exploiting software vulnerabilities to targeting supply chains, social engineering, AI, and more. Here are some of the top threats used today:

1. **Ransomware**: Various variants and gangs target businesses of all sizes, and Ransomware-as-a-Service makes it easy for less sophisticated attackers to use sophisticated tools.

2. **Data Theft and Extortion**: Attackers steal sensitive data and then threaten to publish or sell the data if a ransom isn't paid. This is often paired with a ransomware attack and can be used as a second form of ransom on a single attack.

3. **Credential Theft**: Phishing emails and malware are built to steal credentials, including passwords, as well as MFA tokens, allowing criminals to gain access to accounts or sell the credentials on the dark web.

4. **Business Email Compromise (BEC)**: Increasingly advanced attacks that are designed to trick employees into giving out sensitive information or transferring money to fraudulent accounts.

5. **Supply Chain Attacks**: When a cybercriminal compromises a third-party vendor, and then the hacker

uses the third party's network to gain access to a larger more valuable target.

6. **Living off the Land Binaries (LOLBins)**: Attackers use legitimate and often typical tools and files that are present on a computer or network. Since the tools are legitimate, detecting malicious activities is very difficult.

7. **Remote Monitoring and Management (RMM) Tool Abuse**: Legitimate Remote Monitoring and Management tools, typically used by IT departments and IT providers, are exploited by attackers to either get access to a network, maintain access to a network, or deploy malicious software to a target network.

8. **Cloud Storage Exploitation**: Hackers use cloud storage services such as OneDrive, Google Drive, and more to steal data or deliver malware, using the trust already placed in these platforms.

9. **Nation-State Sponsored Attacks**: State-backed groups that target organizations to steal sensitive information and disrupt critical operations; these groups often leverage advanced persistent threats (APTs).

10. **AI-Driven Attacks**: AI tools, including generative AI such as ChatGPT, have made it much easier for criminals to create advanced cyberattacks, add automation, and avoid common spelling and grammatical mistakes in phishing emails that previously made the emails far easier to identify.

Additionally, AI models themselves are increasingly targeted as a way to steal sensitive data or inject malware.

11. **IoT Device Vulnerabilities**: The explosion of IoT devices has introduced new vulnerabilities, considering the lack of security on these devices, causing significant growth in these attacks.

12. **Vulnerable Drivers and Firmware**: Attackers target and exploit vulnerabilities in device drivers and firmware to gain low-level access to the system; this can allow attackers to bypass traditional security measures like antivirus and antimalware software.

13. **Social Engineering**: Tricking and manipulating individuals into providing confidential information or taking risky action through phishing, vishing (voice phishing), or smishing (SMS or text-based phishing).

14. **Zero-Day Exploits**: Cybercriminals, including Nation-States and APTs, sometimes identify otherwise unknown vulnerabilities in software or firmware. These vulnerabilities are then used to launch cyberattacks, which can go undetected.

15. **Distributed Denial-of-Service (DDoS) Attacks**: These are cyberattacks that aim to overwhelm a system or network's resources and cause a disruption. IoT devices are increasingly being exploited because of their limited security and being used to create a botnet or a large

group of devices that can be used together to launch a DDoS Attack.

ADAPTING TO THE NEW NORMAL

Not only is today's business environment fast-paced and ever-evolving, but the threats are the same, and cyberattacks are increasingly complex, forcing businesses to continually adapt. Cybersecurity and risk management are no longer nice to have; they are a must-have, no longer an afterthought or checkbox. They must be a core function of your strategic planning, as they will help ensure operational resilience. The stakes are as high as they've ever been, and it's not the house's money on the table; the money belongs to the businesses and organizations that drive our economy. Complacency isn't an option; the cost is far too high, and action is the only option.

Businesses must make cybersecurity and cyber-risk management a priority; this means investing in comprehensive security measures, including continuous monitoring, employee security awareness training, regular cyber-risk assessments, incident response plans, robust backup and disaster recovery, vulnerability management, cyber-policies management, and more. While our focus is on businesses, this impacts everyone, so everyone should be a part of the solution.

The evolution of cyber threats is a reminder to us all that change is the only thing we can count on, so we need to be prepared, we need to be vigilant, we need to be proactive, and we need to be resilient. We can do this, but we need to do it together. As things change and threats evolve, we must evolve as well.

Nick Stevens
Heroic Technologies

NICK STEVENS grew up in Silicon Valley, where he caught both the entrepreneurial spirit and the technology bug. His affinity for business and technology is shared with his close friends, many of whom have started, run, and sold successful technology businesses. After graduating from college with a business degree, Nick began his career in technology as an IT support technician, helping businesses manage their computer networks, software, and websites.

Nick's focus is on small to medium enterprises, primarily in the legal, accounting, construction, and technology industries. In 2011, Nick checked two major items off his bucket list: he married the love of his life and co-founded Heroic Technologies, which at the time was a boutique IT consulting firm specializing in law firms in Silicon Valley. While continuing to hone his expertise in technology, Nick relocated to Portland, Oregon, where he opened Heroic Technologies' second office.

Fast forward to today, Nick and his team have built Heroic Technologies into a leading Managed Cybersecurity, IT, and Cyber Compliance firm. With offices in the Portland, Oregon Metro area and the San Jose, California Metro area, Heroic secures and supports organizations across California, Oregon, and Washington, protecting their employees and assets worldwide.

Nick is passionate about balancing people, processes, and technology to help his clients stay productive and secure, no matter where they are. He and his team specialize in Cyber Risk Assessments to identify potential risks within an organization's technology systems, processes, and how employees interact with them. Recognizing that people are a key factor in an organization's cybersecurity, Nick strongly advocates for Security Awareness training to help employees identify and report threats.

Recently, Nick has begun consulting on AI solutions to boost operational efficiency and enhance threat detection while ensuring compliance and proactive risk management.

To learn more or contact our team go to:
https://books.heroictec.com/technology-legislation-is-coming

SCAN THE QR CODE:

TWO
EMERGING CYBERSECURITY STANDARDS
WHAT TO EXPECT

Author: Christopher Lemish
Company: Sniper Watch

THE LANDSCAPE: HERE'S WHAT WE'RE UP AGAINST

IN AN ERA where digital transformation accelerates at an unprecedented pace, the security of our cyber landscapes has never been more critical. Imagine this: a single cyberattack costs the global economy over $1 trillion annually, and the number of data breaches continues to rise exponentially, year after year. With cyber threats evolving at lightning speed, the need for robust cybersecurity standards is imperative and more pressing than ever.

Since 2016, let's highlight the financial impact damage on businesses from a global scale. (Source: *Parachute, Cybercrime Magazine, Statista*)

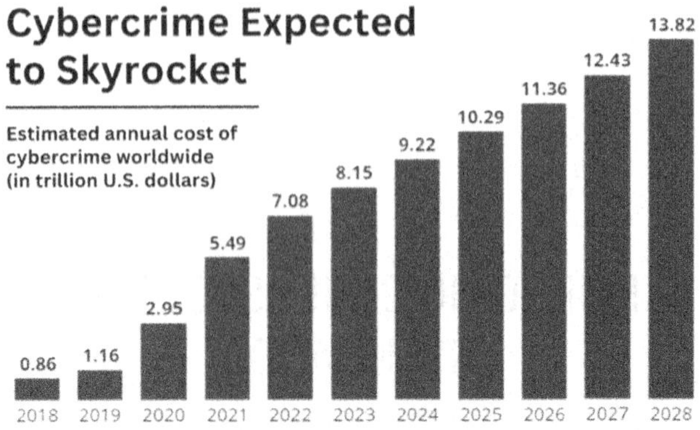

- **2016**: Cyberattacks cost the global economy approximately $450 billion.
- **2017**: The costs rose to approximately $600 billion.
- **2018**: The figure increased further to approximately $860 billion.
- **2019**: Cybercrime costs were estimated at $1.16 trillion.
- **2020**: The cost surged to $3 trillion.
- **2021**: Cybercrime damages were projected to reach $6 trillion.
- **2022**: The estimated cost grew to $7.08 trillion.
- **2023**: The global economic impact of cyberattacks was approximately $8.15 trillion.
- **2024**: The costs are expected to reach $9.22 trillion.
- **2025:** The annual cost of cybercrime is expected to hit $10.29 trillion
- **2026**: Cybercrime costs are estimated to reach $11.36 trillion
- **2027**: The annual cybercrime costs are expected to increase to $12.43 trillion
- **2028**: Costs are expected to reach $13.82 trillion

Now, let's calculate the rate of increase year over year:

- **2017**: 33.33%
- **2018**: 33.33%
- **2019**: 34.88%
- **2020**: *158.62%
- **2021**: *100.0%
- **2022**: 18.00%
- **2023**: 15.11%
- **2024**: 13.13%
- **2025**: 11.61%
- 2026: 10.40%
- 2027: 09.42%
- 2028: 11.18%

In summary, it is undeniable that the financial damage caused by cyberattacks increases every year in the range of 9%-50%, and the expectation approaching 2028 and beyond, is that this trend will continue.

THINGS HAVE CHANGED

Heraclitus, the ancient Greek philosopher, said, "The only thing that is constant is change." His philosophy means that we live in an ever-changing universe, that everything is in a state of flux, and that stability is merely an illusion.

Through the Ages, Man has lived through the Paleolithic Era (Old Stone Age), the Mesolithic Era (Middle Stone Age), the Neolithic Era (New Stone Age) to the Agrarian Era (Agricultural Revolution), the Industrial Era (Industrial Revolution) introducing mechanization, urbanization, and transportation systems to where we are today in the Digital Era (Information Age) comprised of the internet, computing, information accessibility, automation & globalization to where we are now embarking upon in the Post Digital Era of Artificial Intelligence, the Internet of Things (IoT), Biotechnology, and massive gains in Generative AI and Quantum Computing. In order to know where we're going, we have to know from where we've come. If we let history be our greatest teacher, it explains that through each period, it is incumbent upon us to anticipate and welcome change as it is imminent. If we understand the concepts and systems that have been aggregated and amalgamated over time, we can predict the strategies and tactics required to adapt and prepare us for future events.

THE BEGINNING

Security is like fashion. It is never complete. It is dynamic, and the landscape is continuously changing. We are entering into a new paradigm shift, a new conceptual model vastly different from traditional technology to which many have been accustomed. Imagine building a city. Each expert is bringing their own tools and concepts to build their own buildings, transportation points, parks, and shops. It is disparate, disjointed, and unharmonious. There are arguments between the experts because "their way" is the "right way" based on how they've learned to build. This is how things operate today in which not everyone understands the "big picture," and most Subject Matter Experts are architecting and deploying as they see fit toward a mission-less destination.

Consider the alternative, where a specific set of like-minded experts come together to prepare, plan, design, implement, operate, and optimize an ecosystem of how things need to function together to support the people system. Zero Trust, combined with a Defense in Depth strategy, is the foundational element in how a well-designed, thought leadership-focused cybersecurity business strategy framework is constructed. It is a mindset and movement towards a context and attribute-based identity and access control ecosystem. Regulators, government, and insurance providers are impacting the mandate for an Information Security Program. Those who adopt this ideology early are likely to be the disruptors in their business much faster than their competition.

Complacency is the silent killer. A common theme amongst small and medium-sized businesses is that if the enterprise hasn't been attacked, why would the organization spend time and money investing in its security portfolio when it "seems" to be working fine for years? The lack of an attack is considered a performance indicator, which is fundamentally wrong. Past breaches analyzed in our After-Action Review (AAR) reveal that lower attack counts mean that the organization did not have adequate visibility and detection functions to expose observables, or indications of compromise, thus appearing as "everything is fine" at the surface level. Organizational leaders are provided a "false sense of security" by their appointed technical and security advisors when, in fact, significant artifacts and advanced persistent threats (APTs) existed under the covers, evading traditional security defenses before the targeted attack was launched.

WHAT IS ZERO TRUST?

The overall construct of Zero Trust is knowing *who* can access *what*, keeping technical business contextual policy alignment in mind, across each transaction, while employing least privileged access to data. COVID-19 introduced new complexities into technical business operations where teams needed to quickly shift to supporting their remote workforce. The inception of borderless networks to support work from home applied stress to remote access VPN infrastructure and injected scalability concerns. The concept of work from home has changed how organizations provide access with a "work from anywhere, access anything" structure. Traditionally, organizations perceived internal assets as "trusted," but the work landscape has broadened, and the attack surface has commensurately widened.

Zero Trust has been around for many years but is only now starting to make its way into the boardrooms because things need to change, and the traditional ways of doing things have weakened in effectiveness and efficacy. The slow adoption of the concept of Zero Trust has been because of the lack of a change in mindset shift, architectural and technological challenges, lack of technical skill sets, and the lack of process maturity. As we've indicated earlier, attacks on businesses have increased exponentially over the last ten years. Technology and security have traditionally been a reactive defense to events that have done damage or caused disruption. Contrary to the reactive approach, is the more effective and efficient method of a proactive risk mitigation strategy driven by risk and asset protection.

So, what is Zero Trust, really? The major vendors have different ways of achieving Zero Trust that are tailored to their respective products and solutions. From Illumio, Crowdstrike, Cisco, Palo Alto, Zscaler, Forcepoint, Cloudflare, Microsoft to HashiCorp, each has slightly varying concepts, but there is a common theme. The top five principles are:

1. Continuous visibility
2. Context-based access
3. Segmentation
4. Breach assumption
5. Implicit trust removal

We have coined the concept and believe that Zero Trust is a "*Guilty Until Proven Innocent*" methodology while maintaining ease of use in how users interact with their daily activities. Some people believe that Zero Trust is designed to make users' lives harder, but in reality, it is fundamentally rooted in making access easier by allowing users to connect from any device, at any time, and from any location, thus increasing availability to perform job tasks. Many implementers of technology solutions are hyper-focused on their particular solution and features, and the technology could appear rigid. At the heart of the strategy is the end-user experience. We must reverse engineer and work ourselves backward to put ourselves in the shoes of the business unit or persons who are required to achieve their goals or solve their industry challenges. The business must generate revenue, and we cannot allow technology to inhibit these transactions by restricting workflows and impeding user performance. Rather, we must enhance the user efforts.

There must be a delicate balance between security and the overall experience of how the user interacts with their applications and workflows. As an example, as a user and device are onboarded onto a network, it might go through a series of checks in authentication and authorization to determine that they are who they say they are. We have the ability to perform these posturing and vulnerability checks quickly, but during the time that this validation is taking place, we don't want to prohibit the user from other application workload access, so before a user is granted appropriate level access to a corporate network or resource, they are provided with Internet Access Only until their identity has been proven through a series of attribute-based control verification mechanisms, thus not impeding other workflows allowing them to become more productive.

Zero trust is not a product. It is not a vendor coming in and telling you that they will implement Zero Trust because it cannot be sold. It cannot be achieved by implementing expensive hardware or the best hardware available. Zero Trust is a concept, an ideology that encompasses a comprehensive approach to strategic risk mitigation. The reason why many have difficulty implementing this strategy is that it requires various technical solutions to come together to achieve the desired outcomes.

For years, technology has operated in silos, but now business risk and technology have become so integrative that multiple solutions need to complement each other to achieve the protection required in modern business today. Zero Trust is like Aristotle's philosophical work in Metaphysics, "*The whole is greater than the sum of its parts.*" Each solution on its own can help you in your journey toward achieving Zero Trust, but each individual solution is not a silver bullet in accomplishing Zero Trust. When a multi-layered defense approach merges with a strategy where security solutions work together, you develop an ecosystem of detection and visibility that is far greater than any single solution.

Remember, you could have the best technology or hardware in the world, but you must have the skilled personnel to create and build the right policies along with implementing effective processes combined with constant nurturing and continuous monitoring. Otherwise, the offering is still a product and not a comprehensive solution.

Imagine for a moment, a magnetized center core, similar to the earth's core, and all of these siloed technology puzzle piece shapes being activated, attracted, and pulled together to fit in perfect harmony, working together into an overall solution. This is how Zero Trust is constructed in the next evolution of modern business technology.

The Cybersecurity and Infrastructure Security Agency (CISA) has its Zero Trust Maturity Model Version 2, which provides a roadmap for organizations to implement Zero Trust architecture. There are five main pillars of Zero Trust:

1. Identity
2. Devices
3. Networks
4. Applications and Workloads
5. Data

The three capabilities that are applied across all pillars to ensure consistency are Visibility and Analytics, Automation and Orchestration, and Governance.

Zero Trust is a panoptic security ecosystem focused on context-based (attribute-driven) access control designed to minimize uncertainty, ensuring least privileged access to data on a transactional request-by-request basis. Segmentation is a core concept within Zero Trust, and this idea carries us into the macro and microsegmentation levels of identity and access. Application workloads are being designed at a microservice architecture level. Natively, we deploy security on workloads that detect and auto-remediate attempts to install software that is out of compliance, but then furthermore, if an approved, compliant software application is allowed and due to some external stimuli or user interaction, causes malware to detonate and the application starts to communicate or make requests to another irrelevant application, or attempts a Command and Control (CnC) to exfiltrate data to external infrastructure, our API-driven microprocess detection will stop that connection from taking place, alert our SecOps team to be investigated and threat researched.

We are still in the infancy of this design model, and security will exponentially become enhanced in the next ten years. It is my opinion that with the advent of Artificial Intelligence and Machine

Learning, combined with quantum computing processors, the network and security architectures will be gauged with metrics in which intention-based security or Intent will become the standard. The onboarding of devices and the algorithms and microprocesses, ports, protocols, and scans that occur behind the scenes provide context into the attributes or the profiling of the device, such as operating system, make, model, serial numbers, MAC addresses, which determine the OUI (Organizationally Unique Identifier), security patch levels, Syslog, SNMP, NetFlow, Performance Monitoring, Telemetry, and MIB (Management Information Base) values that determine a certainty factor of that device identity. However, the use of cloud-based intelligence engines that can compute hundreds of millions of attributes—before and after actions and the history of previous behavior to determine intent—provides deeper insight into the predictive actions of devices and users, creating behavior-based security.

Imagine a human walking down the sidewalk, passing by your home, and admiring the architecture of the roofline from different angles. They take out a camera or smartphone and snap a few pictures from different vantage points. One might glean that they appreciate your new remodeling of the home, while another form of intention might be that they like the color of your roof shingle. At the same time, another perception could be more malicious in nature. They are collecting data points of vulnerabilities or weaknesses around your home to be sent to a team of thieves preparing to rob and steal your personal belongings once everyone leaves for work one day. As technology gets better, it won't be enough to collect attribute information like the type of camera or height, weight, and hair color of the person, but the data warehousing of information and other behavioral attributes and activities known before and after this incident, or "transaction" would determine malintent based on previously determined and subsequent records and activities. The artifact collection history and top key indicators would form the basis of intent for this action to better determine a benign or malignant action.

Conversely, cybercriminals will adopt and integrate AI into their daily activities as well. AI can be an advantage and a disadvantage. Fire can heat your home and burn it to the ground if misused. Water can keep us hydrated, but it can also cause flooding and damage to cities. AI can serve as a tool in our protection suite, as well as be a tool to enhance deepfakes, bend and distort the truth, and execute misinformation.

GOING DEEPER

Zero Trust is the building of an "Information Security" program. It is a security model that authenticates a subject, providing least privileged access to a data object within the enterprise inventory scope. The access is dynamically generated and validated at the time of requested access based on the identity, location, risk profile, governance, and compliance of the requested source. A centralized logical policy engine that ingests asset information from multiple source collections and provides an access verdict that executes effective user, network, and application segmentation. Continuous verification and monitoring to detect contextual behavior changes upon each request are perpetually evaluated to ensure secure access control to data.

Some vendor solutions claim that Zero Trust can be achieved in software only. Zero Trust is an ecosystem that is balanced between hardware and software. Wireless is provided to endpoints only after hardware connections are made first. You can't pick out your kitchen cabinets without first having a strong basement foundation to support your first floor. Each technology starts with a strong foundation.

As devices are onboarded to an enterprise network, either through wired, wireless, or VPN, the endpoint should not be allowed to communicate on any ports or protocols in which to function until it proves its identity. The only protocols allowed are the basic functionality to allow the device to onboard via DHCP/DNS. The policy server is engaged, and the endpoint is forced to undergo a strict and consistent authentication and authorization process. The first check is to ensure a Public Key Infrastructure (PKI) construct is validated in which the domain joined machine has the proper computer and user certificate. An installed certificate is like a driver's license provided by a DMV. A trusted source that everyone agrees on, highly regulated and monitored, the DMV provides a driver's license—the certificate—which is difficult to spoof.

The next security check is to validate that the computer and user belong to the correct Organizational Units (OUs) in Active Directory (AD). As we check attribute-based parameters, such as whether the authentication process is an 802.1X function and whether the device is a Windows 11 computer, along with compound conditions, we have the ability to assess the posture of every device that onboards. We ensure that each endpoint computer has the correct Endpoint Detection & Response (EDR) or Extended Detection & Response (XDR), formerly Antivirus, installed. Furthermore, we have the ability to agnostically check for any security program and the corporate-compliant version approved by the organization through a validation called application collection.

As mentioned, we design security to enhance the end-user's experience, not impede it. For example, we allow the authorized EDR/XDR and plus or minus two revisions, which would be allowable based on compliance to onboard the network.

As an example, let's say a corporate executive is traveling and has not been able to remediate or install, or have been pushed the latest software revisions for that EDR/XDR. We do not want to deny the executive from connecting to internal resources if he or she were 1-2 revisions off the corporate standard. If more than two revisions, which might be a violation of corporate policy, there might be Common Vulnerabilities and Exposures (CVEs) associated with an outdated version of software; therefore, the executive would need to onboard and undergo remediation, whether manual or through automation to receive the latest updates. All of these transactions occur within milliseconds, and once a successful authentication and authorization lifecycle has occurred, based on the location, device, and identity permissions of the user, they would receive a Dynamic Access Control List (DACL) via an Authorization Profile, which would extend least privileged access to a given network resource. If the user changes their condition, such as their location or access media (i.e., wired to VPN), a Change of Authority, or CoA, would be triggered from the network ecosystem to inform the policy server of an imminent change, and the user and device will be pushed a new DACL to access that network resource, which might be a more restrictive form of access control to maintain adequate security in accordance to data loss prevention (DLP). To further enhance the user experience and minimize downtime, while the authentication and authorization checks are processing, the user will have an Internet-Only DACL in which they can access cloud services until their identity is proven in order to be approved for RFC1918, or internal resources.

Furthering Zero Trust, macro and microsegmentation is an advanced feature in which the user's identity and job function can only access network resources or an application workload with reduced permission sets, which also might be further postured by type of device, time of day, duration, or geographic region. For example, just because a brain surgeon and a pediatrician both hold the title of doctor doesn't mean the brain surgeon should be allowed access to children's medical records. Similarly, in the corporate world, IT personnel should not be able to access human resources' employee records on a particular floor. Think of security group tags as identifiers, similar to a corporate badge. For example, if an IT staff member attempts to swipe their badge for access to an IT Datacenter MDF or IDF closet on the 5th floor, access would be granted. However, if they try to swipe their badge for the opposite wing HR or Payroll room on the same floor, access would be denied. This allows the identifier to go beyond network boundaries. Both employee types can access the same floor, but they have specific privileges to specific locations.

Threat-centric network access control (TC-NAC) is similar to posture in the onboarding process, but we go one step further in securing both the user and device. During the authentication and authorization process, the policy server, behind the scenes, checks with a Vulnerability Management system to scan the device for any Common Vulnerabilities and Exposures (CVEs). If the response indicates a Common Vulnerability Scoring System (CVSS) score above a configured threshold, the endpoint and user are subjected to remediation. This process enforces policy, consistency, and automation to ensure that only authorized and secure devices are allowed, maintaining a clean bill of health.

Lastly, another important feature is Anomalous Endpoint Detection. This feature is like an advanced DHCP Snooping. Once a device is authenticated and authorized, the policy system learns about the MAC address or the device's unique identifier and has profiled the device that it is a tablet, a Windows, or a macOS device. The system has learned and profiled the attribute criteria, and if the MAC address were to be spoofed, which is not difficult to execute, and the system sees the onboarding of this device, the policy enforcement would deny access because it has already been seen and authorized as a Windows 11 computer, but now is functioning with an OUI MAC address of an HP Printer Jet Direct card. An alert would be created, a log initialized, and access would be automatically denied, further enhancing and maintaining security so as not to enable or further the possibility of a cyberattack.

CONCLUSION

In today's complex and ever-evolving threat landscape, the Zero Trust model is no longer just an option but a necessity for organizations of all sizes. By adopting a "never trust, always verify" mindset, businesses can reduce the attack surface, protect sensitive data, and strengthen their security posture. As the traditional network perimeter continues to dissolve, the Zero Trust architecture empowers organizations to stay agile while defending against modern threats. Ultimately, Zero Trust is not just a framework—it's a paradigm shift that secures digital transformation, ensuring that trust is built, earned, and continuously maintained.

Christopher Lemish
Sniper Watch

CHRISTOPHER LEMISH is an MCSE/CCIE-level engineer and a recognized expert in cybersecurity with over 20 years of industry experience. With a proven track record of designing and implementing complex, multi-layered network and security architectures, Christopher has successfully collaborated with global organizations of all sizes.

From protecting critical infrastructure to building highly integrative and scalable security frameworks, Christopher is dedicated to simplifying digital transformation while delivering business resiliency, recovery, and continuity solutions. His work addresses the evolving and demanding needs of today's digital on-premise and hybrid multi-cloud landscapes.

To learn more or contact our team go to:
https://www.sniperwatch.com/
company/about-us

SCAN THE QR CODE:

THREE

THE CONSEQUENCES OF GETTING HACKED

Author: Jonathan Fitzgerald
Company: GDS Technology

IN TODAY'S DIGITAL AGE, the threat of cybercrime is a looming concern for business owners. One of the most significant consequences of getting hacked is the financial impact it can have on your business. When a cybercriminal gains unauthorized access to your systems, they can cause significant financial damage through various means, such as stealing sensitive financial information, conducting fraudulent transactions, or demanding ransom payments to restore access to your data.

FINANCIAL CONSEQUENCES OF GETTING HACKED

The financial consequences of getting hacked can be devastating for businesses of all sizes. In addition to the direct financial losses incurred from stolen funds or data, there are also indirect costs to consider, such as the expenses associated with investigating the breach, notifying customers, and implementing security measures to prevent future attacks. These costs can quickly add up and have a lasting impact on your bottom line.

Furthermore, reputation damage from being hacked can also have a significant financial impact on your business. Customers may lose trust in your company if their personal information is compromised, leading to a loss of business and potential lawsuits. The damage to your brand's reputation can take years to repair and may result in long-term financial repercussions.

Another financial consequence of getting hacked is the data breach itself. In addition to the direct financial losses from stolen data, there may also be regulatory fines and legal fees to contend with. Depending on the nature of the breach and the industry you operate in, these costs can be substantial and may further strain your business's finances.

In conclusion, the financial consequences of getting hacked are far-reaching and can have a lasting impact on your business operations. It is essential for business owners to prioritize cybersecurity measures to protect their company's financial health and reputation. Investing in cybersecurity insurance and implementing robust security protocols can help mitigate the financial risks associated with cybercrime and ensure the long-term success of your business.

DISRUPTION TO DAILY OPERATIONS

One of the most immediate impacts of a cyberattack is the interruption of normal business operations. When critical systems are compromised or shut down, employees may be unable to access important files, communicate with colleagues, or complete essential tasks. This can lead to missed deadlines, lost revenue, and frustrated customers. In some cases, businesses may even be forced to temporarily cease operations while they work to restore their systems and data.

Financial consequences are also a major concern when daily operations are disrupted by a cyberattack. Not only can the cost of repairing and restoring systems be significant, but businesses may also suffer financial losses due to downtime, lost sales, and damaged reputation. Additionally, the time and resources required to address the aftermath of a cyberattack can divert attention and resources away from revenue-generating activities, further impacting the bottom line.

Reputational damage is another consequence of getting hacked that can disrupt daily operations. Customers, partners, and stakeholders may lose trust in a business that has been the victim of a cyberattack, leading to a decline in sales, partnerships, and opportunities. Rebuilding a damaged reputation can take time and effort, further prolonging the disruption to daily operations and hindering business growth.

Data breach consequences can also disrupt daily operations, as businesses may be required to notify customers, regulators, and other parties of the breach and take steps to mitigate further damage. This can be a time-consuming and resource-intensive process that can impact employee morale and productivity. Additionally, the loss of sensitive data can have long-term consequences for a business, including legal liabilities, regulatory fines, and damage to brand reputation.

The disruption to daily operations caused by a cyberattack can have far-reaching consequences for a business. From financial losses and reputational damage to data breach implications and cybersecurity insurance implications, the impact on business operations can be significant and long-lasting. Business owners must be prepared to address these consequences proactively and implement measures to prevent and mitigate the effects of cybercrime on their daily operations.

LOSS OF PRODUCTIVITY

When a business is hacked, employees may not be able to access critical systems or data, leading to downtime and decreased efficiency. This can have a significant impact on the bottom line, as every minute of lost productivity translates to lost revenue.

Furthermore, the time and resources required to restore systems and data after a cyberattack can be substantial. This means that employees may be spending more time dealing with the aftermath of the attack rather than focusing on their regular tasks. As a result, business operations may be disrupted, leading to further losses in productivity.

In addition to the direct impact on productivity, there are also indirect consequences of a cyberattack that can affect a company's ability to operate efficiently. For example, if customer data is compromised, the business may lose the trust of its customers, leading to a decline in sales and revenue. This reputational damage can have long-lasting effects on the company's ability to attract and retain customers.

Data breaches can also result in legal and regulatory consequences that can further impact a company's productivity. For example, if sensitive customer information is exposed in a breach, the company may face fines and penalties for failing to protect that data. This can divert resources away from core business activities and lead to additional losses in productivity.

Overall, the loss of productivity that results from a cyberattack can have far-reaching implications for a business. It is important for business owners to understand the potential consequences of getting hacked and take proactive steps to protect their company from cyber threats. This includes investing in cybersecurity measures, training employees on best practices, and considering cybersecurity insurance to mitigate the financial impact of a potential attack.

LEGAL AND REGULATORY IMPLICATIONS

The consequences of getting hacked extend beyond just financial losses. Business owners must also consider the reputational damage that comes with a data breach. Customers may lose trust in the company's ability to protect their sensitive information, leading to a decline in sales and a damaged brand image. Rebuilding trust with customers and stakeholders can be a long and arduous process, requiring a significant investment of time and resources.

Data breach consequences can also have a lasting impact on business operations. In addition to the immediate financial losses from the attack, companies may face disruptions in their day-to-day activities as they work to recover and secure their systems. This can lead to delays in production, missed deadlines, and decreased productivity, all of which can further impact the bottom line.

For business owners looking to mitigate the financial impact of a cyberattack, cybersecurity insurance can provide some relief. However, it is important to carefully review and understand the terms of the policy to ensure adequate coverage. Some policies may have limitations or exclusions that could leave the company vulnerable to additional costs in the event of a data breach.

Ultimately, the recovery costs associated with a cyberattack can be substantial.

From legal fees and regulatory fines to reputational damage and operational disruptions, the consequences of getting hacked can have far-reaching implications for a business. By understanding the legal and regulatory landscape surrounding cybersecurity, business owners can better prepare themselves for the potential fallout of a cyberattack and take proactive measures to protect their company's assets and reputation.

RECOVERY COSTS ASSOCIATED WITH GETTING HACKED

• Cost of investigating the breach

In the world of cybercrime, one of the most significant costs that business owners face is the cost of investigating a breach. When a company experiences a data breach, it is crucial to conduct a thorough investigation to determine the extent of the damage and identify the vulnerabilities that led to the breach. This process can be time-consuming and expensive, as it may require hiring external cybersecurity experts to assist with the investigation.

The financial consequences of investigating a breach can be substantial. Not only do business owners need to cover the costs of hiring cybersecurity experts, but they may also need to invest in new security measures to prevent future breaches. Additionally, the downtime caused by the investigation can result in lost revenue and productivity, further adding to the financial burden.

Beyond the financial costs, there are also reputational damage consequences to consider. When a company is hacked, it can damage its reputation with customers, partners, and stakeholders. This loss of trust can have long-lasting implications for the business, potentially leading to a loss of customers and revenue.

The consequences of investigating a data breach can also be severe. In addition to the direct financial costs of the investigation, business owners may also face legal and regulatory penalties if customer data is compromised. These penalties can further add to the overall cost of the breach and can have a lasting impact on the business's bottom line. It is essential for business owners to take proactive steps to protect their data and prevent cybercrime before it happens. Investing in cybersecurity measures and cybersecurity insurance can help mitigate the risks and minimize the potential costs of a data breach.

- **Cost of notifying customers and stakeholders**

The cost of notifying customers and stakeholders can vary depending on the size of the breach and the number of individuals or organizations impacted. In some cases, businesses may need to hire a third-party communications firm to assist with the process, adding to the overall expense. Additionally, there may be legal requirements that dictate how and when notifications must be made, further increasing the cost and complexity of the situation.

Beyond the financial implications, reputational considerations should also be taken into account. How a business handles the notification process can have a lasting impact on its reputation and customer trust. Failing to communicate effectively and transparently can lead to further damage to the company's image and credibility, potentially resulting in lost business and negative publicity.

Furthermore, data breach consequences can extend beyond just the initial notification process. Businesses may face lawsuits, regulatory fines, and other legal actions as a result of a breach, further adding to the financial burden. In addition, business operations are affected—downtime, lost productivity, and decreased customer confidence, all contributing to the overall cost of a cyberattack.

In light of these potential consequences, business owners must ensure they have adequate cybersecurity measures in place to prevent breaches from occurring in the first place. Investing in cybersecurity insurance can also provide some financial protection in the event of a breach. Ultimately, the cost of notifying customers and stakeholders is just one of the many expenses that businesses must consider when it comes to the aftermath of getting hacked.

• Cost of implementing new security measures

Implementing new security measures can be a costly endeavor for businesses, but the consequences of not doing so can be even more detrimental. The cost of implementing new security measures can vary depending on the size and complexity of the business, as well as the specific measures being implemented. From investing in robust cybersecurity software to hiring IT professionals to monitor and maintain systems, the financial investment required can add up quickly.

One of the most significant financial consequences of getting hacked is the potential loss of sensitive data. In addition to the cost of recovering lost data, businesses may also face fines and legal fees associated with data breaches. Furthermore, the damage to a company's reputation can have long-lasting effects on customer trust and loyalty, ultimately impacting sales and revenue.

Data breach consequences can extend beyond financial losses to include the loss of customer trust and loyalty.

Once a company's reputation has been tarnished by a data breach, it can be challenging to rebuild trust with customers. This can lead to a decrease in sales and revenue, as well as potential legal repercussions from affected customers.

The impact of getting hacked on business operations can be significant, with downtime and disruptions affecting productivity and revenue. In addition to the cost of recovering from a cyberattack, businesses may also need to invest in new security measures to prevent future breaches. Cybersecurity insurance can help mitigate some of these costs, but it is essential for businesses to carefully review their policies and ensure they have adequate coverage.

Recovery costs associated with getting hacked can include everything from IT repairs and system upgrades to legal fees and customer compensation. By investing in proactive security measures, businesses can minimize the risk of getting hacked and reduce the potential financial and reputational consequences. Ultimately, the cost of implementing new security measures is a small price to pay compared to the potentially devastating consequences of not doing so.

CYBERSECURITY INSURANCE IMPLICATIONS OF GETTING HACKED

Cybersecurity insurance is an essential tool for businesses looking to protect themselves from the financial consequences of getting hacked. There are several types of cybersecurity insurance policies available, each offering different levels of coverage and protection. Understanding the different types of cybersecurity insurance can help business owners make informed decisions about their cybersecurity risk management strategies.

One common type of cybersecurity insurance is data breach insurance, which covers the costs associated with a data breach, including notifying affected customers, providing credit monitoring services, and legal expenses. Data breach insurance can help businesses mitigate the financial and reputational damage from a cyberattack, as well as comply with data breach notification laws.

Another type of cybersecurity insurance is business interruption insurance, which covers the costs of lost revenue and extra expenses incurred as a result of a cyberattack. Business interruption insurance can help businesses recover from the impact on their operations and minimize the disruption to their business.

Reputational damage insurance is another important type of cybersecurity insurance that covers the costs of repairing a business's reputation after a cyberattack. Reputational damage can have long-lasting consequences for a business, including loss of customers and revenue. Reputational damage insurance can help businesses rebuild trust with their customers and stakeholders.

Cyber extortion insurance is a type of cybersecurity insurance that covers the costs of responding to a ransomware attack or other cyber-extortion threats. Cyber extortion insurance can help businesses negotiate with cybercriminals and recover their data without paying exorbitant ransom fees. Understanding the different types of cybersecurity insurance can help business owners protect their businesses from the financial and operational consequences of getting hacked.

COVERAGE OPTIONS AND LIMITATIONS

One of the key limitations of cybersecurity insurance is the coverage options available. Different policies may offer varying levels of coverage for different types of cyberattacks. For example, some policies may only cover data breaches, while others may also cover ransomware attacks or denial of service attacks. It is important for business owners to carefully review their policy and understand what is covered and what is not.

Getting hacked can result in substantial financial consequences, including costs associated with data recovery, business interruption, and legal fees. Cybersecurity insurance can help offset these costs, but it is important for business owners to have a thorough understanding of their policy limits. In some cases, coverage may be capped at a certain dollar amount, which may not fully cover the costs associated with a cyberattack.

Reputational damage from getting hacked can also have long-lasting consequences for a business. Customers may lose trust in a company that has been hacked, leading to a loss of business and damage to the brand's reputation. Cybersecurity insurance may not cover the costs associated with rebuilding trust with customers or repairing the damage to a company's reputation.

Data breach consequences of getting hacked can be severe, including potential fines and penalties for failing to protect customer data. Cybersecurity insurance may help cover some of these costs, but it is important for business owners to understand the limitations of their coverage. Recovery costs associated with getting hacked can also be significant, including costs for forensic investigations, data restoration, and legal fees. Business owners should carefully review their cybersecurity insurance policy to ensure they have the appropriate coverage in place to protect their business from the consequences of getting hacked.

CLAIMS PROCESS AND REQUIREMENTS

In the unfortunate event that your business falls victim to a cyberattack, it is crucial to understand the claims process and requirements for seeking compensation for the damages incurred. This process can be complex and time-consuming, but with the right information and guidance, you can navigate it effectively.

When filing a claim for financial losses resulting from a cyberattack, it is essential to provide detailed documentation of the incident, including evidence of the breach, the impact on your business operations, and the financial consequences suffered. This documentation will help support your claim and increase the likelihood of a successful outcome.

Reputational damage is another significant consequence of getting hacked, as customers may lose trust in your business and take their business elsewhere. When filing a claim for reputational damage, it is crucial to demonstrate the impact on customer trust and loyalty, as well as any efforts made to restore your reputation.

Data breach consequences can be severe, as sensitive information may be compromised and result in legal ramifications. When filing a claim for data breach consequences, you must provide evidence of the breach, the data compromised, and any regulatory fines or penalties imposed as a result. This information will help substantiate your claim and ensure you receive the compensation you deserve.

When filing a claim for the impact on business operations, it is essential to document the extent of the disruption, the financial losses incurred, and any costs associated with recovering from the attack. This information will help support your claim and ensure you receive adequate compensation for your losses.

Jonathan Fitzgerald
GDS Technology

JONATHAN FITZGERALD is the managing partner of GDS Technology. Locally owned GDS Technology offers IT and cyber security solutions to small and medium businesses in Atlanta and provides co-managed and technology services along the East Coast and Midwest. With the increased number of employees working remotely and associated cyber security threats, GDS focuses on providing companies with zero-trust security and virtual desktop infrastructure in the cloud. Their service portfolio also includes managed IT services, co-managed services, cyber security, data backup, compliance, and consulting.

Jonathan attributes the Scouting America program for instilling in him many of the values he possesses and uses today to guide GDS and its employees. His association with Scouting America goes back over 20 years. In 2001, he earned the highest attainable Eagle Scout rank. In 2003, he was awarded the Order of the Arrow – Vigil Honor, a high mark of distinction for demonstrating exceptional service, personal effort, and unselfish interest.

Jonathan has been in the IT business since 2001. He initially honed his skills at a local Atlanta firm that built networks, servers, workstations, and communications hardware. Jonathan then served as president of Choice Data Recovery for five years, creating infrastructure for the data recovery platform, including servers and customizations in BIOS, DOS, and Linux storage.

In 2010, he started a six-year association with Anime Weekend Atlanta, where he provided the technical needs for a 20,000-plus attendee convention, with a support team of six assistant directors and 30 crew members working under him. From 2005 to 2019, he also worked as a freelance consultant to IT firms and businesses, advising them on cyber security, technical design and processes, telecom, and safety.

Jonathan knows that the key to GDS Technology's success is in establishing relationships built on honesty and trust with clients. Every member of the GDS team possesses strong communication skills and top-notch technical skills. Before they make any recommendations, they take the time to learn about their clients' business needs, goals, and priorities. Everyone on the GDS teams knows their success is directly tied to their clients' success. They firmly believe that being an IT provider and cyber security expert means not just reacting to problems but also predicting what may fail and proactively fixing it, avoiding lost time and productivity for their clients.

To supply his clients with specialized expertise without the full-time expense, GDS also offers vCIO (virtual chief information officer), vCSO (virtual chief security office), vCISO (virtual chief information security officer), and vCTO (virtual chief technical officer) services.

In 2021, Jonathan stepped back into volunteering in the scouting program, where he teaches and serves the local community. He also works with Gwinnett County Public Schools and local colleges, providing feedback to ensure their curriculum continues to match the ever-changing world of IT and cyber security.

To learn more or contact our team go to:
www.gdstech.tech

SCAN THE QR CODE:

FOUR

UNDERSTAND YOUR RISK WITH REGULAR ASSESSMENTS AND MANAGEMENT

Author: Jesse Pearson
Company: JPtheGeek

SUPPOSE a construction company is building a new office complex for a major corporation. This company conducted a risk assessment to identify several key risks that may thwart the success of the construction process, delaying the whole project. For instance, the budget may overrun, the supply can get disrupted, or safety hazards hampering the construction site. The construction manager considered these risks to make sure the project is finished on time without any unnecessary delays.

No matter which industry you operate in, you must systematically identify and evaluate possible risks to your organization's success. Regular risk assessment minimizes potential losses and allows organizations to seize newly emerging growth opportunities. Once you understand your risk, you can set your startup for long-term success. Here are the key elements of effective risk assessment:

- Creating clear-cut policies and procedures for risk assessment
- Developing a risk-aware culture for ongoing risk identification
- Continuously improving and learning from your experiences

Once your organization makes risk management part of its overall strategy and starts using relevant tools for decision-making, it'll open the door to long-term success and sustainability. This is precisely what we will explore in this document. So, let's explore the intricate world of risk assessment.

WHAT IS RISK? EXPLAINING THE CONCEPT

Do you know that almost 7 in 10 organizations don't have comprehensive risk assessment procedures in place? If you want to create an effective risk management strategy, you may start by learning the term "risk" in the world of business. Risk means the sort of impact uncertainty has on your goals. How likely is it that you won't be able to achieve your objectives? Anything that hampers your organization's ability to realize its goal is the "risk."

TYPES OF BUSINESS RISK

Some major types of risk in the world of business include the following:

- Financial: This includes fraud, high interest rates, inability to meet obligations, etc.

- Operational: Includes system failure, legal disputes, supply chain disruptions, etc.
- Strategic: Your competitors launch a better product, or customers change their preferences
- Compliance: Your organization failed to meet regulatory compliance, incurring fines, penalties, and lawsuits
- Reputational: Negative online reviews or cybersecurity breaches

Next, we'll explore how conducting regular risk assessments can help you overcome these issues.

WHY IS IT IMPORTANT TO CONDUCT REGULAR RISK ASSESSMENTS?

How will you conduct a risk assessment if you don't have a clear idea of what this concept entails? Let's help you understand the idea behind risk assessment. What is it, and how is it done? We'll explain the proper way to assess your risks in the next section. For now, let's briefly describe the process. Suppose you own a small startup, and you wish to evaluate the factors threatening your business success. Here's what you will do next:

- You will identify the hazards facing your small business
- You will explore how likely each threat is to happen
- Then, you will implement proper control measures to mitigate these risks
- After that, you'll create documentation of findings to craft an effective risk assessment strategy
- In the end, you will keep reviewing this strategy and updating it as needed

For instance, small businesses account for more than 40% of cybersecurity threats every year. If you are ready to face these threats, then you can proactively mitigate them and save your organization a decent amount. Risk assessment also helps you allocate more resources to thwart risks most likely to happen; it stops you from wasting resources or the least realistic risks. That's how the subtle art of risk assessment will help you make well-informed decisions.

HOW TO CONDUCT RISK ASSESSMENT?

How do you conduct a risk assessment and find out what sort of threats your organization is facing? Almost all state/federal regulatory authorities will tell you to conduct a risk assessment at least every year. This process is important to stay on top of the latest threats your organization might face.

5 CRUCIAL STEPS IN RISK ASSESSMENT

- **Risk Identification**

In the first step, you'll use tools and techniques like SWOT analysis, FMEA, bowtie methods, etc., to see which business threats are most likely to prevent you from realizing your objectives. You should create a comprehensive list of risks—ranging from the most likely to the least probable—and stay on top of your risk assessment efforts.

Confused about what these terms mean? Don't fret; we'll explain seven tools/techniques in the last part of this section.

- **Risk Analysis**

After identifying these risks, you'll have to assess their likelihood. How likely are they to affect your goals and objectives? You can use two methods to analyze these risks and the intensity of their impact:

- Qualitative: Risk matrices (explained later in Tools and Techniques)
- Quantitative: Probability distributions (may require programmers)

- **Risk Evaluation**

You can then prioritize risks based on their likelihood and impact. After analyzing their intensity, you can create different categories of business risks. The first category will include the most probable ones. Then, the second category will contain the second-most likeliest risk. The final category will include the threats with the least likelihood. That's how you can focus more on the risks that require immediate attention.

- **Risk Treatment**

The fourth step involves determining the best way to manage business threats. You will discuss different strategies to take care of looming risks. For instance, you may transfer risks, change objectives, mitigate risks through controls, and—most importantly—accept risks if the management decides that these risks are within the organization's risk appetite.

- **Tracking and Reviewing**

Lastly, you shouldn't forget that risk assessment is an ongoing process. Your organization must monitor risks and make sure that their risk mitigation strategies are working fine. If you start realizing that some techniques aren't as effective as before, you can replace them with new strategies.

TOOLS AND TECHNIQUES

As promised, here are some of the major tools/techniques used to assess business risk:

- Risk matrices: Our brain processes pictures 60,000 times faster than words. That's how it's beneficial to create a visual representation of business risks. That's how you can inform your investors and stakeholders about them.
- Risk registers: These comprehensive documents identify and track the status of different threats along with their key details. These details might include the impact of risk, mitigation strategies, assigned ownership, etc.
- FMEA: Failure Mode and Effects Analysis (FMEA). Can identify flaws in your services as well as failure models. Once you identify these flaws, you can see business threats coming miles away.
- Decision trees: It's always great to create hierarchies. Decision trees give you a structured way to visualize the potential outcomes of different risk mitigation options. That's how you can see which risk management strategy would be better.

- Bowtie method: This amazing model shows how risks, the factors causing them, and the effects brought by these risks are related to each other. That's how you can steer clear of high-impact, low-probability business risks.
- Root cause analysis: You'll use this technique to identify the main reason why the risk emerged in the first place. You'll go beyond merely addressing symptoms and try to find out the root causes behind this business threat.
- SWOT analysis: It stands for "strengths, weaknesses, opportunities, and threats." It's a fantastic way to identify not only the risks facing your business but also the major opportunities in front of your company.

Now, we can close the chapter on risk assessment and learn how to manage it efficiently.

RISK MANAGEMENT FRAMEWORKS: A BRIEF OVERVIEW

When it comes to assessing and managing risks, you need to create a set of rules and guidelines to make sure that everyone in your organization can easily identify these threats. That's what the concept of risk management frameworks (RMFs) entails. These frameworks can alternatively be described as structured approaches toward identifying and thwarting organizational threats. That's why the key elements of any RMF must include these:

- Identifying risks caused by different factors
- Assessing how likely a risk is to happen and what would be the fallout
- How can you control or completely eliminate these risks?

- Policies for tracking these risks and monitoring the efficacy of mitigation strategies
- Establishing oversight responsibilities and a risk management culture

POPULAR RISK MANAGEMENT FRAMEWORKS

We can't go into the details of different RMFs, so here's a brief overview of two popular frameworks:

• ISO 31000

The International Organization for Standardization developed this framework in 2009. You may use it no matter the size of your organization. The flexibility and adaptability of this framework make it unique in the world of risk management frameworks.

Pros	Cons
It's a practical framework that gives you detailed guidance on proper risk management	It simply isn't as popular as the COSO ERM risk management framework
Business owners and risk professionals can make sense of it very easily	It doesn't have detailed guidelines on how to apply the framework properly
It focuses not only on risk events but also on the fallout of uncertainty	The brand recognition of COSO ERM is another aspect lacking in ISO 31000

• **COSO ERM**

On the other hand, this one is published by the Committee of Sponsoring Organizations of the Treadway Commission. COSO ERM takes a more holistic, enterprise-wide approach to risk management. That's probably why the financial services industry loves it.

Pros	Cons
It's more popular and recognized than the ISO 31000 risk management framework	Since it's a complex, layered framework, implementing it can be a chore
You can implement it throughout your business with the proper guidance provided	It has intricate concepts and definitions, unlike the easy-to-understand ISO 31000 framework
It explains everything about responsibilities and reporting along the chain of command	It focuses more on analyzing risk events rather than the consequences of uncertainty

CHOOSING THE PERFECT FRAMEWORK

Are you still confused? Do you want to know the correct way to choose a risk management framework? These simple steps will help you choose the perfect framework:

• **Evaluate Your Needs**

What does your organization need in a risk management framework? You should take into account your organization's size and resources. Look into what sort of risk you have to manage. Check major industry regulations and compliance requirements as well.

• Compare Popular Frameworks

Next, you'll evaluate the popular RMFs to find the best one. That's why we shared the pros and cons of two RMFs. Here are every major framework's strong points:

1. IS 31000: Gives you easy-to-comprehend principles and guidelines for risk management
2. COSO ERM: Better if you want an organization-wide approach toward risk management
3. NIST: This framework is ideal for tech companies facing cybersecurity issues

• Refine Your Selection Criteria

Which key factors are you looking for in a risk management framework? You must refine your selection criteria before implementing an enterprise-wide risk management framework. For instance, it's better to consider the following factors:

- How would you incorporate the best practices?
- What sort of supporting materials and tools are available?
- Do you have access to industry-related guidance and applicability?
- What sort of KPIs and performance metrics can you easily include here?
- Does this framework provide organization-wide risk management coverage?

- **Make Flexibility a Priority**

Keep in mind that an ideal RMF shouldn't only meet your organization's current needs but also stay toe-to-toe as your company expands with time. That's because business expansion comes with new risks and evolved threats. So, your risk management framework must be adaptable.

- **Talk to Risk Professionals**

When it comes to complex or highly regulated industries, business owners must get guidance from risk management consultants or advisors. They're called risk professionals (or, as the industry would like to describe them, risk management specialists). Risk management will grow by 5% between the years of 2019 and 2029 as a profession, creating 50,000+ jobs. You can rely on these experts to suggest the perfect RMF for you.

HOW TO INTEGRATE RISK MANAGEMENT INTO ORGANIZATIONAL CULTURE?

Do you want to make risk management an integral part of your organization's DNA? Organizations have become more interested in developing a risk awareness culture since the events of 2007-09. It's mainly due to the fact that 88% of data breaches stem from employees' own mistakes. That's why organizations focus too much on cultivating a culture of risk awareness. The Institute of Risk Culture defines this "risk culture" in these words:

"Risk culture is a term describing the values, beliefs, knowledge, attitudes and understanding about risk shared by a group of people with a common purpose."

In simple words, if employees at all levels understand the importance of risk assessment, then you have created a risk-aware culture in the office. When employees are attuned to potential risks, they are more likely to spot emerging issues early and raise them to management. That's how organizations can address risks before they escalate. Moreover, it creates a sense of shared accountability among employees. They become capable of making well-informed decisions, and organizational resilience becomes a reality.

HOW TO CREATE RISK AWARENESS

You may wonder how to create a culture of risk awareness in your organization. Here's what you can do if you wish to excel at risk assessment and management:

- Assess the current awareness level: Find out what sort of risk awareness your workers have. It can be done via workshops, surveys/quizzes, discussion groups, etc.
- Culture a comprehensive roadmap: This roadmap will have your goals and timelines. You need to have a clear roadmap to monitor your progress later.
- Educate your employees: Explain to your employees what role they can play in mitigating risks. It'd help if you shared TPRM best practices with them for extra awareness.
- Simulate risks and threats: You can deploy occasional drills to check if your employees are more aware of business risks. Experiencing "risk" in a controlled environment may be a great learning opportunity.
- Monitor progress: In the end, ask how much progress you've made. Is your organization highly risk-aware or

not? What else can you do to make your workers aware of business threats?

Case Study: Acme Manufacturing Company

Take the example of the ABCD manufacturing company (we have to keep it anonymous here). This small company produces industrial equipment used in vehicle manufacturing. However, there was a fire in the key manufacturing plant of this company. As a result, the facility was damaged extensively. The supply chain was also affected by this accident. The upper management reviewed their risk management plan; it recommended the following actions:

- Have backup production sites and equipment at secure locations
- Network with other suppliers to procure critical components
- Have detailed procedures in place to restore manufacturing operations

That's how the ABCD manufacturing company activated its business continuity plan quickly. They shifted the production lines to their backup facilities and talked to alternate suppliers for expedited delivery of the replacement parts. They also managed to delay deliveries by keeping their clients in the loop. That's how the company's full production capacity was restored after four months. Their quick response to the fire improved its reputation as customers were impressed by how quickly they overcame the risk. That's how risk management is done!

CONCLUDING REMARKS

Hopefully, you now understand how risk assessment and risk management are done. If your company is working in an uncertain, increasingly complex business environment, you can boost your resilience and adaptability with proper risk management. Ensure that risk evaluation becomes a critical component of not just your project planning but also organizational culture. That's how you can set your company up for long-term success.

Jesse Pearson
JPtheGeek

JESSE PEARSON is the founder and CEO of JPtheGeek, an IT and cybersecurity services company he started over 16 years ago. With more than 20 years of experience in the IT industry, Jesse brings deep technical expertise to his role. Under his leadership, JPtheGeek has evolved to provide customers with the services they need to secure their networks and keep their businesses running smoothly.

JPtheGeek prides itself on exceptional customer service, offering fast response and resolution times with a personal touch. Jesse is passionate about helping clients use technology efficiently while staying protected from cyber threats.

When not at the office, Jesse enjoys spending time with his wife, Susanne, and their two kids, Oliver and Brielle. With Jesse at the helm, JPtheGeek continues to deliver outstanding service and peace of mind to its growing customer base.

To learn more or contact our team go to:
https://jpthegeek.com

SCAN THE QR CODE:

FIVE
THE IMPORTANCE OF A GOOD CYBERSECURITY EMPLOYEE AWARENESS TRAINING PROGRAM

Author: Jay Lewis
Company: 5th Gear Technology Concepts, Inc.

EMPLOYEES PLAY a crucial role in the company's overall cybersecurity strategy. It is easy to make the case that training is critically important, but how often should employees be trained, and what are some of the benefits? Employees should be continuously trained in the evolving threat landscape. Such training can be beneficial to protect sensitive information and reduce the risk of successful attacks. A good program can also improve the incident response time and remediation, reduce the risk of fines and lawsuits, protect the company's or professional's reputation, and reduce the total cost downtime associated with these attacks. Businesses that empower their employees to take an active role with cybersecurity knowledge can help to build a strong human firewall.

Misconceptions often lead small businesses to overlook the importance of cybersecurity awareness training, leaving them vulnerable to various cyber threats. Some of the common misconceptions or excuses we see are:

1. Believing cybersecurity is solely a technology issue.
2. Lacking knowledge or resources on how to implement such training.
3. Assuming criminals only target large corporations.
4. Underestimating the financial and reputational impact of a data breach.
5. Disregarding potential fines from regulatory agencies.
6. Overconfidence on what cybersecurity insurance will cover.

Cybersecurity is not entirely a technological issue. In fact, the National Cybersecurity Alliance (NCA) addresses this in their list of *8 Biggest Small Business Cybersecurity Misconceptions,* "It's a widespread belief that cybersecurity is a tech issue for the geeks to worry about. In fact, most cyberattacks occur through social engineering, where a criminal infiltrates a system through your people and processes. This could involve an employee unwittingly clicking a link in a phishing email, or a vendor being impersonated and sending you a fake invoice. Very few attacks involve the brute-force cracking of an account (assuming the password is strong and unique, that is). Cybersecurity encompasses not just technology, but also the people and processes within an organization. Human error and negligence pose significant threats. Employees who click on malicious links, use weak passwords, or inadvertently share sensitive information can compromise the security of your entire business. Prioritize building a culture of awareness and responsibility among your staff."[1]

Small to Medium Businesses (SMBs) often turn to a Managed Security Service Provider (MSSP) like 5th Gear Technology Concepts, based in Nashville, Tennessee. Jay Lewis, Co-owner and Chief Business Development Officer says, "Small businesses simply do not have the knowledge or the resources to implement an effective employee awareness training program. We have affordable solutions in place to train single proprietors up to companies with thousands of employees. Also, we often hear small business owners claim that hackers are not interested in attacking their small business. However, both published statistics and our own professional experience clearly show that small businesses are indeed frequent targets of cyber-attacks."

Dr. Jane LeClair, currently the President of the Washington Center for Cybersecurity Research & Development, told the United States House of Representatives Committee on Small Business, "SMBs have resource constraints and often ignore cybersecurity in favor of day-to-day operations or other financial needs. Yet SMBs remain a gateway to gain access to clients, business partners, donors, and contractors working with the SMB . . . a backdoor into many large organizations. These organizations frequently lack the knowledge needed to develop and implement a cyber policy or the expertise to develop a response strategy. Surprisingly, 96% of the attacks on SMBs were fundamentally basic attacks. SMBs need employees trained in networking, operating systems, and multiple layers of security."[2]

---

1. "8 Biggest Small Business Cybersecurity Misconceptions," National Cybersecurity Alliance, December 22, 2023, https://staysafeonline.org/resources/8-biggest-small-business-cybersecurity-misconceptions/. Accessed July 4, 2024.
2. Dr. Jane LeClair, "Statement for the Record: Dr. Jane Dr. Jane LeClair, "State-

According to the SEC, referring to attacks on large businesses like eBay, JP Morgan, Home Depot, and Target, “Since the popular press tends to focus on attacks, like these, that target the largest firms, it can be easy to overlook the fact that SMBs are at even greater risk and are far more vulnerable once they are victimized. In fact, for every high-profile breach, there are many more threats to confidential data held by local businesses.”[3] The SEC also says, “...SMBs remain cybercriminals' favorite target. In fact, approximately 75 percent of all spear-phishing scams in June were directed at SMBs, with the very smallest companies—those with 250 employees or fewer—bearing the majority of those attacks.”

In an IBM study in 2020, compromised employee accounts were the most expensive root cause of breaches, costing an average of $3.86 million dollars per breach across all industries (healthcare, financial, pharmaceuticals, energy, industrial, technology, professional services, etc.).[4] In IBM's 2023 report, it showed that the cost average has risen to $4.45 million.[5] Healthcare continues to have the highest average cost at $10.93 million.

---

ment for the Record: Dr. Jane LeClair, Chief Operating Officer, National Cybersecurity Institute at Excelsior College Before the United States House of Representatives Committee on Small Business, Small Business, Big Threat: Protecting Small Businesses from Cyber Attacks, April 22, 2015," U.S. House of Representatives, April 22, 2015.

3. Luis Aguilar, "II. Cybercrime Represents a Very Real, and Very Serious Threat to SMBs," SEC.Gov, October 19, 2015,https://www.sec.gov/newsroom/speeches-statements/cybersecurity-challenges-small-midsize-businesses#_edn4. Accessed July 4, 2024.

4. "IBM Report: Compromised Employee Accounts Led to Most Expensive Data Breaches Over Past Year," IBM Newsroom, July 29, 2020,https://newsroom.ibm.com/2020-07-29-IBM-Report-Compromised-Employee-Accounts-Led-to-Most-Expensive-Data-Breaches-Over-Past-Year. Accessed July 3, 2024.

5. *Cost of a Data Breach Report 2023*, 1st ed. (IBM, 2023), 13.

In professions where there is legal oversight from regulatory agencies, an employee training program is not an option; it is mandatory, and the consequences can be severe. Federal, State, and Professional agencies are empowered to impose civil and/or criminal penalties, and 20% of organizations that experienced a data breach paid $250,000 or more in fines.[6] In most cases, a small business cannot withstand this type of penalty in conjunction with lost sales, reputational damage, and expenses related to proper public notifications. This is why a whopping 60% of SMBs who experienced a cyberattack went out of business within six months of the attack.

Many industries have regulatory requirements that mandate cybersecurity awareness training for employees. These regulatory industries recognize that the human factor plays a significant role in the protection of data that goes beyond technical solutions. Some examples include healthcare, with the Health Insurance Portability and Accountability Act (HIPAA) that mandates safeguards for protecting patient data. The financial services industry has the Gramm-Leach-Bliley Act (GLBA) and the Payment Card Industry Data Security Standard (PCI DSS), which require security measures for financial and payment card information. Although the regulations differ for each sector, they all share common requirements:

- Conduct regular cybersecurity risk assessments (CSRAs): help to identify vulnerabilities, evaluate potential threats, and maintain a strong defense. These should be conducted at least annually, but some industries have specific requirements on how often risk assessments should be conducted.

6. *Cost of a Data Breach Report 2023*, 1st ed. (IBM, 2023), 39.

- Provide employee cybersecurity awareness training: Most industries, as well as cybersecurity insurance companies, require documentation of employee training. This must be ongoing, not a one-time occurrence. The frequency can be tailored to your company's environment based on its complexity, the sensitivity of data handled by employees, the current threat landscape, or from the results of the CSRA.
- Implementation of security controls regarding data handling and privacy protections:
    - Administrative Controls: policies and procedures, security training, and incident response planning.
    - Technical Controls: Access control via MFA (Multi-Factor Authentication), encryption of data at rest and in transit, network security (firewalls, intrusion detection/prevention systems), endpoint security, and application security.
    - Physical Controls: secure facilities (security cameras, access controls, security personnel, etc.).
    - Monitoring and Detection Controls: Security Information and Event Management (SIEM), log monitoring, and anomaly detection.
- Have a written incident response plan: should include steps for preparation, swift identification of the breach, containment of the threat, eradication of malware, and system recovery. After an event, there should be a thorough review of findings to learn from the experience and update security measures.

- Report breaches within specified time limits: this varies by industry, as well as federal and state laws. However, a company can be expected to be required to provide detailed information about the nature and impact of the breach, as well as the mitigation efforts. There will also be a requirement to notify individuals affected by the breach, especially those with an elevated risk of harm.

When a breach occurs, it will trigger a formal audit by a third-party digital forensic company, which may be initiated by regulatory agencies that govern the business, the company's insurance provider, or both. If a business is paying for cybersecurity insurance, the insurance company will be naturally concerned about reducing its exposure through finding and enforcing policy exclusions. Some of these exclusions could include negligence, unpatched systems, regulatory non-compliance, or internal threats caused by malicious actions of employees. When a breach occurs, and a company notifies the proper agencies, they will be required to show documentation to prove compliance with laws, regulations, and policies.

A third-party digital forensic firm will be engaged in investigating the breach. Their scope will include examining logs, network traffic, all affected systems, passwords, policy adherence, and other administrative, technical, and physical controls, as previously mentioned. They will identify exclusions, evaluate compliance, and report the findings, including the root cause of the breach. Regulatory agencies and insurance companies then scrutinize the reports to determine if there is justification for fines, penalties, or the coverage limits of the claims. A business that does not make a reasonable effort to follow the requirements of regulations, laws, and policies will suffer more stringent assessments and increased scrutiny on future renewals.

Businesses that do not have a documented and ongoing cybersecurity training program create an avoidable and self-imposed problem.

A third-party forensic investigation can be an expensive burden on a small business if they are found to be negligent or ambivalent. The insurance company may require the business to absorb the full financial burden of the investigation. It is important for a business to review this with their insurance company to determine the anticipated procedures and costs associated with a breach claim. According to ScienceSphere.blog, "When a company experiences a data breach, digital forensics is crucial in identifying the source of the breach and determining the extent of the damage. The cost of digital forensics in such cases can vary depending on the size of the organization and the complexity of the breach. On average, the cost can range from $10,000 to $50,000, including the expenses related to data preservation, analysis, and documentation."[7]

Employees should be trained in multiple ways and on a frequent basis. This training should be documented to comply with both regulatory agencies and cybersecurity insurance requirements. Some examples include:

- Awareness of diverse types of threats, such as malware, social engineering, and phishing.
- Implementation of security best practices, including strong password management and identifying suspicious emails and links.
- Practical skills training with hands-on exercises, allowing employees to apply their security skills in real-world scenarios.

---

7. "Unveiling The Price Tag: How Much Does Digital Forensics Really Cost?" ScienceSphere.Blog, 24 Aug. 2023, www.semaforobares.com/. Accessed 4 Jul. 2024.

- Incident reporting procedures for potential security threats.
- Proper handling, storage, and transmission of sensitive information.
- Following company policies and procedures, as well as understanding regulatory responsibilities.

New threats are always emerging and becoming more sophisticated with the use of artificial intelligence (AI) tools. AI tools are used to scrape data from social media and professional websites on individuals, so cyberattacks are personalized and more likely to be successful. Without a cybersecurity awareness program that is constantly evolving, employees can lack the security awareness needed to optimize their ability to recognize new risks and attack methods.

In conclusion, SMBs typically lack the internal resources to properly educate their own employees, so it is becoming clear why more businesses are turning to Managed Security Service Providers (MSSPs) these days. The risks that companies face are growing more serious all the time. Regulatory agencies are imposing harsher penalties, both financial and legal. Clients are more likely to file lawsuits when things go wrong, and for that matter, let us not forget the damage that can happen to a company's reputation. All of this adds up to a lot of stress for business owners and managers. MSSPs offer expertise in navigating these complex security and compliance issues. By partnering with these specialists, companies can better protect themselves against potential threats and violations. To reduce risk, it is about staying ahead of the curve in an increasingly challenging business environment.

Keeping employees engaged is crucial in ongoing cybersecurity training to ensure they will remain vigilant and informed about the latest threats. Some recommendations are as follows:

1. Real-world scenarios: using studies and examples of actual security incidents to highlight the importance of cybersecurity practices.
2. Regular simulations: using phishing simulations on a non-predictable routine.
3. Leadership involvement: there needs to be a top-down commitment with visible participation in training.
4. Microlearning: weekly videos and quizzes in addition to phishing campaigns.
5. Incentives and rewards: acknowledge and reward employees who do the training and offer incentives like gift cards or paid time off.

Businesses that empower their employees to take an active role with cybersecurity knowledge become an integral part of a company's overall defense against cyber threats. Companies can create a human firewall against cybercriminal gangs. Going back to the quote from the National Cybersecurity Alliance earlier in this chapter, "Employees who click on malicious links, use weak passwords, or inadvertently share sensitive information can compromise the security of your entire business. Prioritize building a culture of awareness and responsibility among your staff."[8]

---

8. National Cybersecurity Alliance, "8 Biggest Small Business Cybersecurity Misconceptions."

Jay Lewis
5th Gear Technology Concepts, Inc.

JAY LEWIS is co-owner and CBDO/CFO at 5th Gear Technology Concepts Inc. in Nashville, Tennessee. He has extensive expertise in cybersecurity and is a retired criminal police investigator with experience in cybersecurity and major crimes.

To learn more or contact our team go to:
https://5thgearconcepts.com/techlegislationbook/

SCAN THE QR CODE:

SIX

PROTECTING MOBILE DEVICES AND REMOTE WORKERS

Author: Eddie Kee

Company: Kee Innovations Inc.

THE LANDSCAPE of cybersecurity technology and remote work has significantly evolved. Many employees, contractors, and business associates are working remotely, away from traditional office settings. While this offers flexibility, it also brings forth a unique set of security challenges that must be addressed. These challenges are as diverse as the remote workers themselves. While some require full access to office resources, including file sharing, printers, and email, others only need secure internet access. Balancing these varying needs is a complex technological challenge for organizations. A well-written remote policy can help with compliance testing.

This chapter explores the importance of securing mobile devices and remote workers and provides a comprehensive guide to best practices, tools, and future trends.

UNDERSTANDING THE THREAT LANDSCAPE

Mobile devices and remote workers are prime targets for cybercriminals. Common threats include malware, phishing attacks, and unsecured Wi-Fi networks. Real-world examples, such as the 2020 Twitter hack, highlight the vulnerabilities and potential consequences of inadequate security measures.

The threat landscape is varied. You need to ensure that all updates are applied to the latest OS version for your specific hardware, whether it's Mac, Windows, Linux, iOS, or Android. Each account and device should have a unique password, and it's best to use different passwords for each. Enable multi-factor authentication to prevent unauthorized access to your accounts. It's important to encrypt your data, which is a built-in feature on many business-class operating systems. Ensure that encryption is enabled on your smartphones and tablets. Additionally, browse the Internet safely by using an updated browser and turning on private or isolated browsing.

1. Dangers of Mobile Device Usage

While convenient, mobile devices pose significant security risks. These include:

Data Leakage: Data leakage occurs when sensitive information is unintentionally exposed through unsecured apps and services. This can happen when apps request excessive permissions or when data is transmitted without encryption. For example, an app that accesses your contacts, location, and messages without a legitimate need can compromise your privacy. To mitigate this risk, users should review app permissions regularly and avoid installing apps from untrusted sources.

Phishing Attacks: Mobile users are often targeted through SMS (text message) and email phishing attacks. These attacks trick users into providing personal information or clicking on malicious links. Phishing messages may appear to come from legitimate sources, such as banks or social media platforms, making them difficult to detect. To protect against phishing, users should be cautious of unsolicited messages, verify the sender's identity, and avoid clicking suspicious links.

Malware: Mobile devices can be infected with malware that steals data or damages the device. Malware can be disguised as legitimate apps or delivered through malicious websites. Once installed, malware can access sensitive information, track user activity, and even take control of the device. To prevent malware infections, users should install antivirus software, keep their operating system and apps updated, and avoid downloading apps from unofficial app stores.

Physical Security: Loss or theft of mobile devices can lead to data breaches. If a device falls into the wrong hands, the data stored on the device can be accessed and misused. To enhance physical security, users should enable device encryption, use strong passwords or biometric authentication, and activate remote wipe capabilities to erase data if the device is lost or stolen.

Juice Jacking: Juice jacking occurs when a mobile device is charged through a compromised USB port, such as those found in public charging stations (Airports, Coffee Shops, Libraries, Colleges, etc.). These ports can be used to install malware or steal data from the device. To avoid juice jacking, users should carry their own chargers and use electrical outlets instead of public USB ports. Alternatively, USB data blockers can be used to prevent data transfer while charging.

Wi-Fi Breach: Connecting to unsecured Wi-Fi networks can expose mobile devices to various security risks. Cybercriminals can intercept data transmitted over these networks, leading to data breaches. To protect against Wi-Fi breaches, users should avoid connecting to public Wi-Fi networks and use a virtual private network (VPN) to encrypt their internet traffic.

Open Port Scanning: Open port scanning involves scanning a device for open ports that can be exploited by cybercriminals. Open ports can provide unauthorized access to the device and its data. To mitigate this risk, users should disable unnecessary services and ports, use firewalls to block unauthorized access, and regularly update their device's security settings.

Connection to Rogue Wi-Fi Networks: Rogue Wi-Fi networks are malicious networks set up by cybercriminals to intercept data transmitted by unsuspecting users. These networks often mimic legitimate networks, making them difficult to identify. When users connect to a rogue Wi-Fi network, their data can be intercepted and stolen. To protect against this threat, users should verify the legitimacy of Wi-Fi networks before connecting and use VPNs to encrypt data in transit. Encryption ensures that even if data is intercepted, it cannot be read without the decryption.

2. Mobile Device Security

Mobile device security is equally as important as the security measures you take on your computer. Securing mobile devices is crucial as they often contain sensitive information and access to corporate networks. Best practices include:

Regular Updates: Keeping the operating system and apps

updated to patch vulnerabilities. Most commercial software on our devices is updated regularly. The updates can be applied automatically, or a manual installation can be requested. If you are leaving on an extended trip, it is best to force a manual update.

Strong, Unique Passwords: Using complex passwords and changing them regularly. A password is something that you know. For example, using a password like "!Sunshine1234" would meet the complexity requirements of most security systems, but it would be easily guessed. Try using a passphrase like "GoingToTheSunnyPark-At2pm!"—this option is more secure. Avoid using a password that needs to be written down to be remembered. Also, consider using a password manager. The password manager will generate a password if requested. The generated password is of the form "Abcde1!-Fghij2*-Klmno3@." While this is secure, most users would need to write it down to use it.

Two-Factor Authentication (2FA): Adding an extra layer of security to prevent unauthorized access.

Encryption: Protecting data at rest and in transit. Encrypting your data is essential!, The functionality is built into many business-class operating systems. Ensure that encryption is enabled on your smartphones and tablets.

Safe Browsing and Downloads: Avoiding suspicious links and apps. Browse the Internet safely by using a known updated browser and enabling private or isolated browsing.

Tools like mobile device management (MDM) software can help enforce these practices.

One of the major ways that mobile devices are compromised is via rogue wi-fi networks, as discussed above. Some signs of rogue wi-fi networks include:

Unfamiliar Network Names: Be cautious of Wi-Fi networks with names that are similar but not identical to legitimate networks.

For example, if you see "CoffeeShop_WiFi" and "CoffeeShop_FreeWiFi," the latter could be a rogue network.

No Password Protection: Legitimate networks usually require a password. If a network that should be secure is suddenly open and doesn't require a password, it might be a rogue network.

Duplicate Network Names: If you see multiple networks with the same name, one of them could be a rogue network. Cybercriminals often create networks with the same name as legitimate ones to trick users.

Unusual Network Behavior: If you connect to a network and experience unusual behavior, such as frequent disconnections, slow internet speeds, or unexpected pop-ups, it could be a rogue network.

Suspicious Network Activity: Use network monitoring tools to check for unusual activity. If you notice a high volume of data being sent or received without your knowledge, it could indicate a rogue network.

Precautions to avoid attacks via rogue Wi-Fi include:

Verify Network Names: Always verify the network name with the establishment providing the Wi-Fi. For example, ask the staff at a coffee shop for the correct network name.

Use a VPN: A Virtual Private Network (VPN) encrypts your internet traffic, making it difficult for cybercriminals to intercept your data, even if you connect to a rogue network.

Enable Firewall and Security Software: Ensure your device's firewall is enabled and use up-to-date security software to protect against potential threats.

Avoid Automatic Connections: Disable the automatic connection feature on your device to prevent it from connecting to unknown networks without your consent.

Check for HTTPS: When browsing the internet, ensure that the websites you visit use HTTPS. This indicates that the data transmitted between your device and the website is encrypted.

3. Remote Worker Security

Remote work introduces unique security challenges like unsecured home networks and increased phishing attempts. Best practices include:

Secure Wi-Fi Connections: Using WPA3 encryption and strong passwords. For example, a laptop connected to public Wi-Fi at an airport would have posed significant security risks in the past. However, today, we have software that can monitor and block all incoming requests, providing safe and secure usage of public Wi-Fi. Additionally, carrying a mobile hotspot connected to reputable mobile providers can provide secure internet access on the go, thanks to built-in firewalls.

VPN Usage: Encrypting internet traffic to protect data.

Regular Data Backups: Ensuring data can be recovered in case of a breach.

Phishing Awareness: Training employees to recognize and avoid phishing attempts.

Other constant threats, such as physical risks like loss, theft, or damage to the computer, also need to be considered. Additional remote access considerations may arise when traveling overseas, necessitating using computers with no access to internal company information or even disposable computers based on the destination. Tools like endpoint protection platforms (EPP) and secure access service edge (SASE) solutions can enhance remote worker security.

4. Creating a Security Culture

A strong security culture is essential for protecting mobile devices and remote workers. This involves:

Training and Awareness: Regular training sessions on security best practices.

Security Policies: Developing and enforcing mobile device and remote work security policies.

Regular Security Audits: Conducting audits to identify and address vulnerabilities.

5. Future Trends in Mobile and Remote Work Security

The rapid advancement of technology continues to reshape the landscape of mobile and remote work security. Two of the most influential emerging technologies are the Internet of Things (IoT) and Artificial Intelligence (AI).

Internet of Things (IoT):

- *Increased Connectivity*: IoT devices, such as smart home gadgets, wearables, and connected office equipment, are becoming ubiquitous. While they enhance productivity and convenience, they also introduce new vulnerabilities. Each connected device is a potential entry point for cyberattacks. For instance, a compromised smart thermostat could be used to gain access to a home network and, subsequently, to corporate data.
- *Data Proliferation*: IoT devices generate vast amounts of data, which can be valuable for businesses and attractive

to cybercriminals. Protecting this data requires robust encryption and secure data management practices.

- *Device Management*: Managing and securing a growing number of IoT devices is a significant challenge. Organizations must implement comprehensive IoT security frameworks that include device authentication, regular firmware updates, and network segmentation to isolate IoT devices from critical systems.

Artificial Intelligence (AI):

- *Enhanced Threat Detection*: AI and machine learning algorithms can analyze vast amounts of data to identify patterns and detect anomalies that may indicate a security threat. This capability allows for faster and more accurate threat detection and response.
- *Automated Security Measures*: AI can automate routine security tasks, such as patch management, threat hunting, and incident response. This reduces the burden on IT teams and ensures that security measures are consistently applied.
- *Predictive Analytics*: AI can predict potential security threats by analyzing historical data and identifying trends. This proactive approach enables organizations to address vulnerabilities before they are exploited.

PREPARING FOR FUTURE SECURITY THREATS IN YOUR WORKPLACE

As technology evolves, so do the tactics of cybercriminals. Preparing for future security threats requires a proactive and adaptive approach. Here are some strategies to consider:

1. Continuous Monitoring and Threat Intelligence

Real-Time Monitoring: Implementing real-time monitoring solutions helps detect and respond to threats as they occur. Security Information and Event Management (SIEM) systems can aggregate and analyze data from various sources to provide a comprehensive view of the security landscape.

Threat Intelligence: Leveraging threat intelligence feeds provides insights into emerging threats and attack vectors. This information can be used to update security measures and stay ahead of cybercriminals.

Zero Trust Architecture: Zero Trust Architecture (ZTA) is a security model that operates on the principle of "never trust, always verify." Unlike traditional security models that assume everything inside an organization's network is trustworthy, Zero Trust assumes that threats can exist inside and outside the network. Therefore, strict verification is required for every user and device attempting to access resources, regardless of location.

Core Principles of Zero Trust Architecture:

- *Least Privilege Access*: Users and devices are granted the minimum level of access necessary to perform their tasks.

- *Implementation*: Access controls are fine-tuned to ensure that users can only access the resources they need. For example, an employee in the marketing department should not have access to financial records.
- *Microsegmentation*: The network is divided into smaller, isolated segments to limit the spread of potential breaches.
- *Implementation*: Each segment is protected with its own security controls. For instance, sensitive data might be stored in a separate segment with stricter access controls than general data.
- *Continuous Monitoring and Validation*: Continuous monitoring of user activity and device health to detect and respond to threats in real-time.
- *Implementation*: Security Information and Event Management (SIEM) systems and Endpoint Detection and Response (EDR) tools are used to monitor and analyze activity across the network.
- *Multi-Factor Authentication (MFA)*: Requiring multiple forms of verification before granting access.
- *Implementation*: Users might need to provide a password and a one-time code sent to their mobile device to access a system.
- *Device Health and Compliance*: Ensuring that devices meet security standards before they are allowed to connect to the network.
- *Implementation*: Devices are checked for compliance with security policies, such as having the latest updates and antivirus software installed.

Security by Design

- *Integrating Security Early*: Security should be integrated into the design and development of systems and applications from the outset. This approach, known as Security by Design, ensures that security considerations are embedded in every stage of the development lifecycle.
- *Regular Security Assessments*: Conducting regular security assessments, including penetration testing and vulnerability scanning, helps identify and address potential weaknesses before they can be exploited.

Employee Training and Awareness:

- *Ongoing Education*: Continuous training and awareness programs are crucial for informing employees about the latest security threats and best practices. Phishing simulations, security workshops, and regular updates on emerging threats can help build a security-conscious workforce.
- *Security Culture*: Fostering a culture of security within the organization encourages employees to take an active role in protecting sensitive information. This includes reporting suspicious activities and adhering to security policies.

Collaboration and Information Sharing:

- *Industry Collaboration*: Collaborating with industry peers, government agencies, and cybersecurity organizations can enhance an organization's ability to defend against threats. Information-sharing initiatives, such as threat intelligence-sharing platforms, enable organizations to learn from each other's experiences and improve their security posture.

- *Public-Private Partnerships*: Engaging in public-private partnerships can provide access to additional resources and expertise. These partnerships can facilitate coordinated responses to large-scale cyber threats and improve overall cybersecurity resilience.

Adaptive Security Strategies:

- *Flexibility and Scalability*: Security strategies must be flexible and scalable to adapt to changing threat landscapes. This includes adopting cloud-based security solutions that can scale with the organization's needs and implementing adaptive security measures that can respond to evolving threats.
- *Resilience and Recovery*: Building resilience into security strategies ensures that organizations can quickly recover from security incidents. This includes having robust incident response plans, regular data backups, and disaster recovery procedures in place.

Organizations can better protect their mobile devices and remote workers from future security threats by staying informed about emerging technologies and adopting proactive security measures. This forward-thinking approach is essential for maintaining the integrity and confidentiality of sensitive information in an increasingly connected world.

CONCLUSION

Throughout this chapter, we have explored various aspects of mobile device and remote worker security:

Understanding the Threat Landscape: Recognizing the common threats and real-world examples of security breaches helps organizations stay vigilant and proactive in their security measures.

Mobile Device Security: Implementing best practices such as regular updates, strong passwords, two-factor authentication, encryption, and safe browsing can significantly reduce the risk of security breaches.

Remote Worker Security: Addressing the unique challenges of remote work by securing Wi-Fi connections, using VPNs, regularly backing up data, and raising phishing awareness is essential for protecting remote workers.

Creating a Security Culture: Fostering a culture of security through training, awareness, and regular audits ensures that employees are equipped to handle security threats effectively.

Future Trends in Mobile and Remote Work Security: Staying informed about emerging technologies like IoT and AI, and preparing for future security threats, helps organizations remain resilient in the face of evolving cyber threats.

THE IMPORTANCE OF A PROACTIVE APPROACH

A proactive approach to security is crucial for mitigating risks and ensuring the safety of mobile devices and remote workers. This involves continuously monitoring for threats, updating security measures, and staying informed about the latest developments in cybersecurity. By adopting a proactive stance, organizations can identify and address vulnerabilities before they are exploited by cybercriminals.

LEVERAGING TECHNOLOGY AND TOOLS

Utilizing the right tools and technologies is essential for enhancing security. Mobile device management (MDM) software, endpoint protection platforms (EPP), secure access service edge (SASE) solutions, and other security tools can help organizations enforce security policies, monitor for threats, and respond to incidents effectively. Investing in these technologies is a critical component of a comprehensive security strategy.

BUILDING A STRONG SECURITY CULTURE

Creating a strong security culture within the organization is vital for ensuring that all employees understand the importance of security and are committed to following best practices. This involves regular training sessions, clear communication of security policies, and fostering an environment where employees feel empowered to report suspicious activities. A strong security culture not only enhances the organization's security posture but also helps build trust with clients and stakeholders.

PREPARING FOR THE FUTURE

The cybersecurity landscape is constantly evolving, and organizations must be prepared to adapt to new threats and challenges. This includes staying informed about emerging technologies, such as IoT and AI, and understanding their potential impact on security. Organizations can stay ahead of cybercriminals and protect their sensitive information by adopting a forward-thinking approach and continuously updating security strategies.

FINAL THOUGHTS

Securing mobile devices and remote workers is a multifaceted challenge that requires a comprehensive and proactive approach. By understanding the threat landscape, implementing best practices, leveraging the right tools, and fostering a strong security culture, organizations can effectively protect their sensitive information and maintain operational integrity. As technology continues to evolve, staying informed and prepared for future security threats will be essential for ensuring the safety and security of mobile devices and remote workers.

By following the guidelines and strategies outlined in this chapter, organizations can create a robust security framework that addresses the unique challenges of mobile and remote work. This not only helps protect sensitive information but also ensures that employees can work securely and efficiently, regardless of their location. The importance of securing mobile devices and remote workers cannot be overstated, and it is a critical component of any organization's overall cybersecurity strategy.

Eddie Kee
Kee Innovations Inc.

EDDIE KEE owns Kee Innovations, a cybersecurity and technology consulting firm that helps organizations leverage state-of-the-art technology to grow their businesses securely. A seasoned professional with over 40 years of experience in information technology and engineering, Eddie's expertise includes consulting in gas and oil measurement, remote telemetry, Exchange server farms, and mainframe system management.

He authored a book titled *Networking Illustrated*, which has been translated into six languages, and has co-authored several other networking books.

In addition to his professional achievements, Eddie is actively involved in the Stark County community. He has served as Consistory President of Grace United Church of Christ in Uniontown, Master of Hoover Lodge 770, and Assistant Scoutmaster of BSA Troop 135 in Uniontown.

In his free time, Eddie enjoys camping and traveling with his wife, Diane, son, Noah, and daughter, Katie.

To learn more or contact our team go to:
https://keeinnovation.com/eddie

SCAN THE QR CODE:

SEVEN
IMPLEMENTING TECHNICAL SAFEGUARDS:
CONTROLLING THE CONTROLLABLE

Author: Andrew Hollister
Company: Simple Tech Innovations, Inc.

IT WAS ABOUT 3:00 in the afternoon when I saw my cell phone ringing. The caller ID popped up. It was the Vice President of a client who notoriously disregarded our advice and often made decisions based on their own calculated ROI.

Calling my cell phone meant one thing. There was a problem. About six months before, I had a meeting with this client about their security "posture"—a fancy way of saying that they weren't taking cybersecurity seriously. In fact, they weren't even considering it. Despite the fact that they stored data that was regulated by state law and had compliance requirements, the owner brushed off our recommendations, saying, "I've done the math, and I think as long as I don't get fined once every six years, then I'll break even on your services so I don't think it will be worth it."

That conversation echoed in my memory as I answered the phone... The VP then proceeded to tell me that she had been contacted by the FBI Cybersecurity team that someone had gained access to their servers and network and that the administrative access was on auction for sale on the Dark Web.

The news itself wasn't surprising. This customer refused to do a Cybersecurity Risk Assessment, so they had no idea what their risk was, and as their service provider (I'm using that term quite loosely), we didn't know either.

My next question: "Do you have Cybersecurity insurance?"

Customer: "No"

Knowing very well that this could mean the complete end of their company, the first thing I offered to do was contact the FBI directly to ensure that this was not a scam and that what she had shared with me was, in fact, true. Within 24 hours, I had confirmation from the FBI that not only was the information true. However, the client incorrectly relayed the information. The information that the FBI was trying to give them was that someone had already purchased admin access to their network and systems. Simply put, they were in trouble.

In many ways, this client's story is not unique. Many businesses, especially small local companies, think, *"It will never happen to me."* They see the big names in the news and believe that big companies are the main target. In reality, small businesses are the low-hanging fruit.

Small businesses are least likely to have invested in cybersecurity measures, they are least likely to have implemented training for their staff, and they are least likely to be following the rules and regulations. But it doesn't stop there. Many times, small businesses think that they are safe because the information they have is stored in the cloud, or maybe they believe they don't actually collect sensitive information to begin with. This false sense of security is usually because there isn't enough education, media coverage, and awareness surrounding the reality of our cyber situation. Because of this, small businesses are prime targets for even the most inexperienced hackers.

In one way, this client is fairly unique compared to others, and that is, the FBI somehow found out about their breach and notified them in a fairly reasonable time. This is the only instance that I have ever personally had where the FBI was able to notify someone that they found they had their credentials actively for sale on the dark web.

Typically speaking, when we do Risk Assessments, we find breached accounts and hacked credentials on the dark web, and sometimes we find systems that are actively hacked and data being exfiltrated out of their networks. But this was the only instance where the FBI caught it and reached out.

So now that the "it won't happen to me" myth is debunked, what do you need to do about it? You need to take control of your computers, mobile devices, network, servers, and all infrastructure. You need to control the controllable.

You know the mantra "The first step to solving a problem is admitting you have one."? In this case, the first step to controlling the controllable is admitting you can control it. You may not know how, or you may not want to, or you may think it isn't necessary, but you do, in fact, have control over what safeguards get put in place on your devices, and you do control the level of protection that is put in place.

At the end of the day, this concept is incredibly important because when you have a breach, law enforcement, your insurance company, your clients, and other third parties will be looking at how much you controlled that was controllable and determining if you controlled an adequate amount, or if you were negligent in what you did not control.

Depending on compliance requirements, data security laws, and your insurance policy, these factors could determine your company's ability to continue doing business. As a decision maker, it could determine whether or not you are open to civil penalties or worse.

Today, technology is moving at a breakneck pace. I would love to discuss specific tools and technologies that can protect your systems today. By the time this book is published, you'll find that half of them could be out of date or obsolete. So, instead, I want to review the implementation of technical safeguards and what they are.

Let's start by explaining what technical safeguards are. In simple terms, technical safeguards are the barriers that you put between yourself, your staff, your network, your data, and hackers who want to ruin your business. Typically speaking, this translates to tools and software. But it could be access control for a door or a physical security system (if you still have a physical office).

One of the biggest mistakes I see in cybersecurity is the assumption that simply purchasing and installing a software package or protection service will ensure proper security and protection.

Unfortunately, I see this just as much with IT companies claiming to provide cybersecurity services or to be specialists in cybersecurity as I see clients who willfully admit they have no knowledge of what to do to protect themselves and struggle to send an email. I won't cover it in this chapter, but it is so important to cut through all the marketing gimmicks and ensure that the service provider you are working with really understands security. IT guys are not security guys. They fix networks and computers but rarely know how to protect them.

The primary way you implement technical safeguards is by first assessing your risk. If you don't understand your risk, you very well may be installing airbags on a bicycle (yes, the type you peddle with your legs) that never leaves your garage while ripping down the highway in a Corvette with no top and no seatbelts going well over 100mph.

If you think of a car, there are many aspects that keep you and the car safe. You have things that alert those around you of your intended actions, like blinkers, brake lights, and reverse lights. There are seat belts, airbags, bumpers, anti-lock breaks, and other safety measures to protect you when an accident happens.

You have security measures such as car alarms, locking doors, and GPS tracking. There are also other features, like blind spot assist, sensors that alert you when you are too close while backing up, and even fail-safes that prevent you from shifting from Drive to Reverse while driving on the highway.

Think of these features as the technical safeguards for your vehicle. They are installed, they are fairly automated but may require some of your intervention, they should be checked regularly to ensure they are performing correctly, and their purpose is to take most decision-making out of being safe and protected.

Imagine if you could only deploy airbags moments before a crash was going to happen, and it required your manual intervention to make it work. This would be a pretty crappy protection, and quite honestly, no one would consider it a safety feature. This is what happens when you implement safeguards that require a manual intervention from a person. Some of the first examples that come to mind are manual backups instead of automated backups. In over 20 years of IT experience, I have yet to have a single client who was doing manual backups of their data able to restore it when they truly needed it. Don't put yourself in a situation where you have to know when someone is going to run a red light, and you need to press the button to deploy the airbags before they hit you.

If we stick with the vehicle analogy, we can understand the role of technical safeguards for our networks, data, and systems.

So, let's look at some of these safeguards and see how they relate to your computer systems.

Your seatbelts and airbags are meant to protect your physical body in the event of a crash. If they are operating properly and properly secured (read that as wearing your seatbelt), they automatically engage when a crash happens. You do not need to have any special interventions. This type of safeguard can be thought of as the modern-day equivalent of endpoint protection. At the time of writing this book (2024), anti-virus is quickly fading and being replaced by Endpoint Detection and Response (EDR), but it very likely will be called something new and flashy tomorrow.

But also consider if you have seatbelts in your car and don't actually put them on. During a collision, they won't protect you, and you will potentially fly right through the windshield of your vehicle. This is the equivalent of buying a software package and not having a professional install it, configure it, and maintain it. You may have it in your vehicle, but if it isn't being used properly, the protections will fail.

Your turn signals, break lights, and horns are the equivalent of monitoring and alerting tools. When something is happening, you are signaling to someone, "I'm breaking," "I will be turning soon," "*Beep, beep*, you are going to hit me!" Your monitoring and alerting tools should be checking your cloud apps, monitoring your physical devices, and checking for irregular behaviors and malicious actions. Then, they should alert your IT team or service provider. But just like a horn. You can beep all night, but if no one is around to hear it, no one will know. So, if you don't have someone monitoring your systems 24x7x365, your alerts can be received too late or not at all, and there will be no response.

Insurance is the easy one, right? Someone backs into your car and does a hit-and-run. You come out and find the front of your vehicle totally crunched. Maybe with no evidence pointing to who hit you at all. If you don't have insurance, you are SOL. You might not even have the money to cover the cost of the repairs, or they may cost more than the value of your car. Maybe you do have insurance, but not collision and comp. So, even though you have insurance, you don't have the right insurance to cover your claim, and you are left in the same position as if you didn't have insurance at all. Or, you have coverage—the right coverage— you pay your deductible, and your car is made whole again after going to the shop.

Ask yourself which scenario your cybersecurity insurance falls in. Do you have any? Are you paying for coverage that isn't truly going to cover a breach? Or are you nestled in the right policy that will make you whole again when you become a victim of a hacker?

Another thing to consider about insurance is that some insurance companies will not insure a vehicle that is missing specific safety features, or they may increase your rate to make up for the risk. Insurance companies definitely won't insure a vehicle that they deem not to be roadworthy. The risk is too high. Whether or not insurance should be considered a technical safeguard can reasonably be debated. But I highly recommend you have cybersecurity insurance, and you likely have car insurance, so I'm including it anyway.

What about the door locks, car alarms, low jack, and other security measures a car has? These are your user accounts coupled with passwords, your authentication systems that ensure you log in from only the place you are authorized to log in from, and the systems that require additional proof that your account or device is what or who you say you are.

Let's not forget about OnStar (I'm sure there are other companies that do this, but they are the most widely known and advertised at this time). If you have OnStar and you get into a crash, your safety measures have deployed, your alerts have gone out, and the OnStar people come on the speaker and call help for you. Shortly after, EMS arrives, and if needed, you are dispatched to healthcare professionals who can care for you. This entire experience is the equivalent of Managed Detection and Response. You have a true team of experts monitoring these systems and responding when one of these systems triggers an alert or detects something suspicious. This team goes into action, locking down the threats and hunting to see what happened.

Now that you have a few analogies to help demystify what technical safeguards are and how they relate to your network and systems, I ask you to consider: Is your IT infrastructure road-worthy and insurable? Is it safe to take a drive on the Internet with your technological vehicle?

You may wonder what safeguards are considered reasonable for your business and what are common safeguards today. Each business and industry has different regulations and compliances that your business will need to adhere to. These requirements will outline a minimum baseline of what your business needs to implement and control.

Your insurance carrier will likely dictate the safeguards as well to keep and maintain coverage. In many cases, additional factors, such as your company's size and annual revenue, will be taken into consideration. But remember: the smaller your business, the more likely you are to be a major target of hackers. So even if you're not required to protect everything. You absolutely should protect as much as you possibly can.

Your best bet to fully know and understand what safeguards you need to implement is to hire a reputable Managed IT Services

Provider who truly understands cybersecurity and the compliances that apply to your business. Have them do a comprehensive Cybersecurity Risk Assessment so you know what you need and what you are missing. You will likely be very surprised with the results. Even if you thought you were really on top of things.

To get you started, I've listed the most common technical safeguards that I see regulations and insurance companies requiring from businesses of all sizes.

Auto-logouts/Screen Timeouts

When you get up and walk away from your computer, does the screen lock and require a password?

Access Control & Least Privilege

You should ensure that your users only have access to the bare minimum to get their job done and nothing more. No user should have access to data, systems, or software that is not directly related to their role and job function. This ensures that in the event of a compromised account, the affected systems are isolated. Only those who need access to sensitive information should have it.

Multi-Factor Authentication

This is a protection on your logins. It starts with something you know (typically your password), and then a second form of authentication. This can be something you have (a smartphone, smart card, USB key) or something you "are," like a fingerprint or facial recognition (typically biometrics).

Password Expiration

This is starting to become obsolete. More and more evidence has shown that forcing users to regularly change their passwords leads to less security, not more. But some things take a long time to change. So, some regulations and compliances still require this.

Password Complexity Requirements

This safeguard ensures that your passwords are at least a certain length and include upper case, lower case, numbers, and special characters.

Disk Encryption

Did you know if someone stole your computer today and your disk isn't encrypted, they can read all of your files without even needing your password? Disk encryption scrambles your files so that if someone takes your hard drive, they cannot access your sensitive data.

VPNs

Virtual Private Networks (VPN) is a way to encrypt the communication between your computer and your office or server. Without a VPN, if you are working from home or in a coffee shop, someone may be able to see the data you are sending and accessing.

Email Threat Protection

Email Threat Protection is a system that receives your emails before they come into your mailbox and checks to see if they are safe. It looks to see if the email is coming from who it says it is coming from, checks to see if it is coming from countries you have approved, and if it has anything malicious in it—bad links, viruses, etc.

Cloud Backups

Cloud backups take your local data and make an encrypted copy in the cloud, so if the device fails (typically a server, but it could be a workstation as well), you can download a copy and restore it.

Email Encryption

Did you know that when you send an email, it can be clear and open for anyone to read? Technology has improved, but as of the writing of this book, you need to actually set up email encryption. It is not on by default and is usually a separate service that you need to purchase through your IT provider or email provider. Much like Disk Encryption, Email Encryption is going to scramble your message so that only the person you sent it to can open and read it.

Endpoint Protection

Traditional antivirus software is essentially obsolete and has been replaced by Endpoint Detection and Response (EDR). However, many people are accustomed to calling it antivirus, so it's easier to describe Endpoint Protection as the modern-day equivalent of antivirus. It is much more advanced because it can do behavioral analysis to see if actions that are happening on your devices are malicious, even if there is no known virus on the computer.

Central Authentication and Unique User Accounts and Passwords

All usernames and passwords should be totally unique. No one in your business should know each other's passwords. I know you are thinking, "But what if I need to access [insert employee]'s email or computer?" There are correct and secure ways to do this without ever needing to know their password. A true IT professional can help you set this up. Your usernames and passwords should authenticate against a server (cloud or local) to ensure they are accurate and correct.

24x7x365 Monitoring and Response

We are seeing this more and more. I wouldn't be surprised if, in the next year or two, it is a base minimum requirement for most, if not all, businesses across the U.S. just to get insurance coverage. This is having a team of true cybersecurity professionals monitor your endpoints, logs, and security software and then respond when something malicious is detected. You should be weary if your IT company says they handle this in-house and extra weary if you have in-house IT doing it because the cost, training, and skill to set up an organization that can truly do this effectively costs millions of dollars to operate.

Patch Management

Patch management is the process of auditing your computers and other devices to see what security updates they are missing and making sure to install them. Much of this is automated, but you will find that updates fail, and they don't install when they said they would. Installing security updates in a timely manner is very important. You could have excellent security software and still leave the door wide open and unlocked by missing important security updates. We have already seen some insurance companies state that your coverage will be reduced if a hack was done on a device that was missing a security update depending on how long ago the update was released (ex: only 50% coverage if the update was released 30 days ago).

Data Loss Prevention

Data loss prevention (DLP) is a way to control what files can leave your network and cloud environment. You can do this by controlling the sensitivity of each file, by globally controlling who can share files externally, or by a combination of these and other filters, such as which devices can access files to ensure that sensitive files do not leave your network.

Audit Logging

Most compliances require some level of logging so that user access can be audited at a later date. These logs typically include the unique user who logged in, what they accessed, when they accessed it, and where they accessed it from. These logs not only help you prove compliance with your policies, but in the event of a breach, they also assist investigators in understanding how the breach occurred and what data was accessed.

Mobile Device Management

Just about everyone has a smartphone, tablet, or other mobile device. Many employees access their work emails from their personal devices. When employees connect their mobile devices to their company email, they are also typically syncing other sensitive company files in the background that they didn't even know about. Mobile Device Management allows you to approve which devices can access company accounts and data, giving you control. It also allows you to control what can be accessed on those devices and enforce specific security policies on those devices, such as forcing lock screens to be enabled, enforcing disk encryption, and other protective measures.

No Daily Driver Admin Accounts

Don't use the administrative account as your day-to-day account or your "daily driver." As the owner of a company, your user account is likely an administrator account. This means if you get hacked, everything in your business will be compromised. This also means your account will be the number one targeted account by hackers. You should create an administrator account that you only use on very rare occasions when you need it and keep your normal account restricted like everyone else's. This way, the number one attacked account is restricted from compromising your entire business. For an extra layer of security, you shouldn't have access to the account at all. Your IT provider should maintain access, and you should request it only on an as-needed basis.

Vulnerability Management

This may sound similar to Patch Management, but it is, in fact, different. Vulnerability Management is scanning your devices, apps, websites, etc. Basically, all of your IT assets (even your printers), seeing what holes exist in your systems, and then checking to see if there are configuration changes, software changes, or other changes that need to be made to protect your systems.

Zero Trust

This one is a bonus because I haven't seen it truly required yet, but it likely will be soon. Many security professionals agree that this is one of the most important controls you can implement in today's threat landscape. However, most IT providers won't want to implement it because it means they will have to work hard. Essentially, this is creating a list of what is allowed (it could be applications, network connections, file access, etc) and blocking everything else. It gives zero trust to everything by default and only allows the few things you allow. As you can imagine, this means you have to be very diligent about updating your list of what is allowed on a regular basis to ensure your computers can run the software you want them to. But it provides excellent security, ensuring only the things you allow are happening.

There are many other technical safeguards that I haven't listed. An entire book could be written about the subject. I remember a time when simply installing anti-virus, cloud backup, and turning on automatic Windows Updates was enough to keep a business secure. Now, you need a specialized team to stay on top of compliance to let you know what controls are required and a team to manage and maintain those controls moving forward.

To recap, I ask you to consider: *Is your IT infrastructure road-worthy and insurable? Is it safe to take a drive on the Internet with your technological vehicle?*

Your homework:

1. Have a Risk Assessment conducted.
2. Review the safeguards and controls missing in your assessment.
3. Implement a service provider who can install and configure those controls.
4. Have that service provider manage, monitor, and maintain those controls.

Andrew Hollister
Simple Tech Innovations, Inc.

ANDREW HOLLISTER is the CEO and Founder of Simple Tech Innovations, a Rochester, NY-based Managed IT Services company. With over 20 years of experience, Andrew is passionate about driving innovation in IT support and cybersecurity, particularly in compliance-driven industries such as healthcare, insurance, and finance.

Renowned for his ability to demystify complex technological solutions, Andrew provides clients across industries with clear, actionable insights. Under his leadership, Simple Tech Innovations has earned multiple awards for exceptional customer service, sustained growth, and outstanding client feedback.

Andrew's influence extends beyond his company—he has delivered seminars on the transformative potential of AI in business, participated in expert panels on delegation, and contributed to various security-related boards. His commitment to helping business leaders safeguard their operations and data from cyber threats has established him as a trusted advisor in the cybersecurity space.

To learn more or contact our team go to:
http://gett3k.com/book_author

SCAN THE QR CODE:

EIGHT
IMPLEMENTING SECURITY POLICIES AND DOCUMENTATION:
ADDRESSING WHAT YOU CAN'T CONTROL

Author: Ben Johnson
Company: CEO of Liberty Technology

IF SOMEONE SAYS that they have cybersecurity figured out, you have met either a liar or a fool. Over the past 20 years, I've learned that being secure isn't only about what you can control but, more importantly, managing what you can't. I've frequently said: "You can spend a fortune on the best security platforms out there, but you can't always prevent human error, malicious insiders, or external threats from compromising your data and systems." Properly implementing security policies and documentation is the best way to protect your information assets and comply with legal and regulatory requirements.

At Liberty Technology, we've seen it all! Small Businesses that think they are "too small" to be a target, to large enterprises that assume they are invincible.

But it's just not true: NO ONE is impervious to hackers. Cybersecurity is the great equalizer, impacting every industry and business size without discrimination. Be it a Fortune 500 company or a tiny family-owned business, the fact is that the risks are very real and completely non-discriminatory.

But here's the catch: despite our best efforts, some threats remain outside our control. It's like trying to forecast the weather—you might have the best meteorologists and the finest technology, but occasionally, Mother Nature throws you a curveball. Similarly, in cybersecurity, there are things we just can't manage. External threats include human error and third-party vulnerabilities, and the list goes on. So, what do we do? Do you give up and just hope for the best? Absolutely not.

The secret is in preparation and resilience. The approach is about using adequate security measures and compiling robust documentation to weather any storm. It's about realizing that even though you can't control everything, you can control how you react. We have spent years helping businesses do just that. Our approach combines common sense with deep knowledge. We make sure our clients aren't just safe. They're ready for what comes next.

Let's break it down: security policy and documentation are like rulebooks for your organization's cybersecurity game. They explain the dos and don'ts, who is in charge, and everything in between. However, unlike a static rulebook, the documents are not set in stone; they are constantly evolving. It will be a dynamic, living, malleable organism because as threats mature, so will the policies and documentation you put into place.

Let's begin with risk assessment and management, the foundation of all security policies. Picture yourself trying to build a structure without understanding the physical geography of the land. You would be signing up for failure. In the same way, a geographical survey provides a map, a rigorous risk assessment gauges conceivable threats, discerns their probability and expected level of damage, and then establishes priorities. Liberty Technology's Compliance as a Service offer employs a risk management framework that dynamically measures and shores your infrastructure against the unknown.

Moreover, it is essential to create comprehensive security protocols. These regulations recognize the hazards previously identified and provide a roadmap for mitigating them. These protocols, from access control policies and data protection guidelines to incident response plans and employee training programs, serve as a blueprint for your organization's cybersecurity strategy. At Liberty Technology, we tailor these policies to align with each client's operational needs, ensuring they're secure and productive.

A user can defeat even the most robust and secure system simply by clicking "OK" (repeatedly). Regular training, phishing simulation exercises, and clear instructions for protecting systems and data can significantly reduce human error. We've seen firsthand how continuous education turns a vulnerable workforce into a formidable first line of defense.

And now, for the elephant in the room—outside factors like external threats, third-party vulnerabilities, and unpredictable human behavior are variables we can neither predict nor manage. What we can do is set up strict vendor management and audit security controls regularly to help mitigate these risks. Liberty Technology's advanced security and monitoring solutions feature robust third-party management that provides all-encompassing security.

One of the most critical and overlooked factors is incident response and recovery. Incidents can and will happen, even with the most excellent plans. It's not if but when. The key is to minimize the exposure and damage that can occur to your organization. An effective incident response plan plus a trusted partner to execute a remediation plan is crucial to minimizing damage and recovering quickly. Our vCISO offering includes a tailored incident response plan to ensure prompt, effective responses to any breach.

Continuous monitoring and revising security posture are essential to avoiding cyber threats. The dynamic nature of the security landscape means that measures also need to be constantly changing. We manage advanced AI-assisted platforms that help keep our clients ahead of the bad guys.

While we can't control every variable, we can control our response. In short, an ounce of prevention is worth a pound of cure. Implementing robust security policies, conducting regular risk assessments and audits, and, most importantly, fostering an environment of constant vigilance can help us confidently navigate the digital landscape. At Liberty Technology, we believe that preparation and resilience are the keys to effective cybersecurity.

PRACTICAL STEPS FOR IMPLEMENTING SECURITY POLICIES

• Risk Assessment and Management

One must first conduct a comprehensive risk assessment to establish a focused security policy. Without identifying the assets to protect and assessing the risk on those assets, it doesn't matter what you do. A risk assessment is like a health check for your IT environment. It provides the contextual guidance required to identify and prioritize potential threats.

At Liberty Technology, we've refined a comprehensive risk management framework to ensure our clients' systems are constantly evaluated and fortified against new threats. The procedure starts by recognizing all critical assets inside the association, going from hardware and software to people and data. Every asset is assessed for its significance to business operations, ensuring nothing falls through the cracks.

Once we identify the organization's assets, we determine the threats. Threats could come from several sources, including cyberattacks such as phishing and ransomware, physical threats like natural disasters, or even unauthorized access by an employee. Moreover, determining the vulnerabilities of each asset is essential. For example, an asset could be vulnerable because of outdated software, lack of encryption, or inadequate access controls.

After identifying threats and vulnerabilities, you must evaluate the potential impact of those threats. This evaluation involves identifying direct and indirect costs such as reputational damage, financial loss, or operational disruption. By using a matrix, we prioritize risks based on their likelihood and impact, focusing first on the high-priority risks. Then, we compile actions to mitigate risks into an action plan, recognize the risks we cannot reduce, and document their probability and impact.

- **Developing Comprehensive Security Policies**

Once risks are recognized, the next logical step is to develop policies that address these threats. These policies serve as your business' playbook for cybersecurity strategy, outlining clear guidelines and procedures to mitigate identified risks and ensure Compliance with industry standards.

Security policies must contain several key elements. Access control policies define who can access what resources, under what conditions, and how this access will be granted and revoked. Data protection guidelines detail how data should be encrypted, stored, and transmitted to protect sensitive information from unauthorized access or breaches. Incident response plans describe the steps for identifying, reporting, and responding to security incidents, including communication processes and recovery actions. You must conduct training programs regularly to inform employees about security threats and best practices.

At Liberty Technology, we customize our policies for each client to match their specific operational needs. These policies ensure our clients are secure while being practical and functional, allowing for smooth integration into their daily business operations.

- **Employee Training and Awareness**

Employee training and awareness are vital aspects of any security strategy and a crucial defense against human error — the primary cause of security breaches. Regular training programs, phishing simulation exercises, and clear communication of security policies are non-negotiable aspects of an overall security plan that dramatically reduce human error.

In a striking example from a county in Georgia, we implemented regular phishing tests as part of employee training sessions. Over six months, the number of employees falling for phishing attempts dropped by 80%. The net result was a much more security-aware organization, with employees far more confident reporting suspicious activities and handling security measures.

At Liberty Technology, we stress the importance of continuous education and have deployed many tools to keep employees informed and on their toes. Frequent bulletins on new threats, "gamified" training methods, and a recognition program for those employees with exceptional security awareness are a few of our strategies.

ADDRESSING THE UNCONTROLLABLE

• External Threats and third-party vendors

External threats and third-party vendors represent factors that are beyond the control of a company and require meticulous attention. Implementing strict vendor management policies and performing frequent security audits can help mitigate these threats. Liberty Technology's Compliance as a Service offering includes robust third-party management practices to increase security dramatically.

The security practices of third-party vendors are critical to your organization's overall security posture because they have access to your sensitive data and systems. It is vital to comprehensively evaluate potential vendors' security measures before signing a contract. Regular security audits confirm adherence to your security standards, and your contractual requirements should detail your security expectations and compliance requirements.

A story that happens too often is when a third party is the source of a breach. A regional healthcare provider, for instance, suffered a data breach stemming from their EMS platform. We helped them strengthen their security posture and updated their supplier management policies. We helped them edit their vendor contracts to require more robust security, established more precise communication channels for timely reporting of incidents, and implemented continuous monitoring of third-party activities. This situation is an excellent example of mitigating an uncontrollable risk—we decreased the risk. If another incident occurs, the provider will have a legal defense and suffer less liability.

- **Incident Response and Recovery**

Despite the best policies and preventive measures, incidents can and will occur. An effective incident response plan is crucial for minimizing damage and recovering quickly. Liberty Technology's vCISO offering includes tailored incident response plans to ensure rapid and effective responses to any breach.

Preparing means to define roles, establish communication, and identify critical systems and data. Detecting and analyzing includes having tools and processes for finding and understanding incidents. Containing, eradicating, and recovering are necessary to limit the damage, eliminate the threats, and get back to normal. Finally, a post-incident review lets you learn lessons and do better next time.

We take several critical steps at Liberty Technology when we begin an incident response.

We use AI-assisted platforms to detect anomalies and potential security incidents automatically so we can respond promptly and take immediate corrective action. We establish clearly defined communication channels to ensure that information is disseminated to all relevant stakeholders efficiently and quickly. Finally, we conduct regular incident response drills, and by reviewing our response procedures, we can ensure that we are prepared to respond effectively to an incident.

- **Continuous Monitoring and Adaptation**

Cyber threats are constantly changing, necessitating real-time monitoring and adapting security policies. With Liberty Technology's managed security services, we can provide the infrastructure necessary for continuous monitoring to ensure that security measures are adapted promptly.

We implemented an adaptive security strategy for a leading national exposition services company, allowing continuous monitoring and responding to new threats. By taking steps such as using Security Information and Event Management (SIEM) tools and Intrusion Detection Systems (IDS) and running vulnerability scanners, we proactively monitor their networks and systems for potential threats, catching them ahead of time and preventing them from causing any significant damage.

Furthermore, by regularly sitting down with our key stakeholders to discuss our policies and engaging them in reviewing them, we ensure they stay current and effective. Similarly, by researching industry trends and staying aware of what new types of threats are emerging, we can help you stay ahead of people who might be trying to infiltrate your network.

CONCLUSION

In the ever-changing realm of cybersecurity, the key to resilience lies in preparation and adaptability. We can't mitigate every risk, but we are responsible for our actions. To create a robust defense against foreseeable threats, we conduct thorough risk assessments, develop comprehensive security policies, and maintain continuous employee training for our clients. Further, we mitigate the uncontrollable—like external threats and third-party vendor vulnerabilities—through our strict management practices and regular audits that further fortify your security posture.

It's just as vital to have effective incident response and recovery plans in place for when a cybersecurity breach happens. They help to minimize "dwell time," the period during which an attacker has taken control and is doing damage, and ensure a rapid recovery. Continuous monitoring and adaptation of security measures keep your organization one step ahead of emerging threats. They will ensure that policies are still relevant and practical.

Liberty Technology has built its reputation by allowing companies to handle such complexities confidently. We provide complete solutions developed with your unique operational environment in mind. We aim to ensure that your security measures are practical and robust.

It is essential to keep in mind that cybersecurity is not something you do once and then never have to worry about again. Threats constantly change, so you must always look for new ways to protect yourself. This variation means updating your security policies regularly, keeping up with the latest threats, and training your employees to spot and combat potential risks. A good defense is proactive, not reactive.

Collaboration is also an essential aspect of supporting security. Attending cybersecurity forums and understanding other perspectives and lessons learned will give you more substantial insights into making your organization more secure. One tip we recommend organizations do is build a trusted network of partners and resources within the sphere they work out of.

Moreover, the use of technological advancements is vital. Security Information and Event Management (SIEM) systems, Intrusion Detection Systems (IDS), and vulnerability scanners are instruments that can provide immediate analysis and monitoring, which can help identify and contain threats before they escalate. Maintaining and updating these tools and policies gives your defenses parity with the threats they expose.

However, it is essential to remember that humans are at the center. The employees who operate an organization's network, applications, and data possess knowledge of the IT environment and its goals, vision, and culture. This knowledge is rarely documented well enough or communicated widely sufficient to enable an AI algorithm to make "good" risk-reducing decisions on behalf of the organization.

So, what does a forward-thinking cybersecurity organization do today? It ensures that everyone knows how important it is for enterprises and employees to understand those exact enterprise control judgments around information risk. Many organizations are launching corporate-wide security threat education to help employees understand when, how, and how successful threat actors are in targeting an organization's data. Participants learn about phishing, spear-phishing, spam, bump-hunting, and other information risk targets.

At the heart of any robust cybersecurity strategy, an organization will find regulatory Compliance, threat intelligence, situational awareness, incident response, and holistic governance. However, once you get past all that operational work, you need to concentrate on the core of your defense-in-depth strategy: your employees. They are the ones operating the network, applications, and data exponentially.

By incorporating these practices into your everyday work, you can defend your resources and lay down a secure way of life throughout your institution. This mindset will prompt attentiveness, responsibility, and a creative response to menaces before they come about—all of which grant a superior security stance as the digital environment worsens.

It's important to recognize that the security measures organizations put in place always have a higher power going beyond procedures and technology put in place. As security methodologies identify threats and weaknesses, understand risk profiles, and generate approaches, there is always uncertainty regarding changing threats and vulnerabilities. This was the case in significant data breaches at companies like Target, Anthem, and Sony, to name a few, which had all the proper security measures in place. However, despite the layered security methods, hackers still found their way in.

If you're ready to put these strategies into action, we've got a ton of resources to help you get started, including free templates and checklists. Go to http://www.libertytech.net/solutions/tools and access these tools today to solidify your cyber defenses. Be ready for anything and design for it because, in the great game of cybersecurity, it's not about avoiding the storm but learning to dance in the rain.

Ben Johnson
CEO of Liberty Technology

BEN JOHNSON, Founder and CEO of Liberty Technology, is a seasoned entrepreneur and cybersecurity expert known for his exceptional ability to launch successful businesses in challenging economic climates, often with limited initial capital. His journey is a testament to his tenacity, adaptability, and deep expertise in both technology and business.

Born and raised in Georgia, Ben earned an Associate of Arts in Psychology from Gordon College, a Bachelor of Arts in English Literature from the University of Georgia, and a Master of Science in Computer Information Systems from Georgia State University. This diverse academic background provided him with a unique perspective that has been invaluable in his entrepreneurial endeavors and equipped him with the technical expertise to navigate the rapidly evolving information technology landscape. Additionally, Ben both attended and later taught at Griffin Technical College, now known as Southern Crescent Technical College.

Despite facing economic downturns, Ben's entrepreneurial spirit remained undeterred. He has a proven ability to identify opportunities and leverage his extensive knowledge to launch and grow businesses. His ventures are driven by innovation and a steadfast commitment to staying ahead in the tech industry. As technology has rapidly advanced, so too have cybersecurity threats, particularly impacting Small and Medium-Sized Businesses (SMBs). Recognizing this critical need, Ben founded Liberty Technology, which has become a beacon of protection for businesses navigating the digital landscape.

Under Ben's leadership, Liberty Technology has established itself as a prominent player in the IT Managed Service Provider sector. The company specializes in helping businesses across multiple industries secure and manage their IT infrastructure while enabling them to drive bottom-line growth. Ben's expertise in cybersecurity has been instrumental in safeguarding clients against evolving threats, earning Liberty Technology a reputation for excellence and reliability. With over two decades of experience, Ben is a dynamic force in the technology and cybersecurity landscape.

One of Ben's guiding principles is encapsulated in the phrase, "We drink our own champagne." This reflects his dedication to rigorously testing and vetting solutions for clients, ensuring they receive the highest quality service and products available.

Beyond his professional accomplishments, Ben finds fulfillment in his role as a devoted husband to his high school sweetheart, Tracy, and a loving father to their daughter, Alexa. His strong family values provide a foundation for his ambitious pursuits and give him a well-rounded perspective on life.

To learn more or contact our team go to:
https://www.libertytech.net/solutions/
cybersecurity-risk-assessment/

SCAN THE QR CODE:

NINE
INCIDENT RESPONSE PLANNING:
WHAT WILL YOU DO WHEN YOU GET HACKED?

Author: Steve Schulenberg
Company: JCS Technologies, Inc.

THE DAY STARTED like any other: wake up, grab a cup of coffee, and scroll through the news briefs. I had just put my son on the bus for school and was tidying up the house a bit when I received the call. You know the call I'm talking about, the one that every business owner and IT department head hates to receive—"We've been hacked!" Right away, I started going through my mental checklist of things we needed to do and in what order they needed to be done. Little did I know how ill-prepared we were for what had just happened.

I was a freelance IT support technician tasked with taking care of basic technology requests for a small business in the industrial solutions industry. The in-house IT manager, Tom, was someone from the accounting department who knew just enough about computers to be dangerous. Business continuity processes were rarely thought of, let alone documented, but backups were both local and online. My responsibilities were limited to "break-fix" as the big projects were handled by Tom.

Upon arrival at the customer site, I started my initial diagnosis. It wasn't long before I determined that their definition of "hacked" meant that a user had clicked on a link in a fraudulent email, and now all his local files had been encrypted. That's right; they were the victim of a Business Email Compromise, or BEC. To make matters worse, because all users had mapped drives that pointed to the same server, all files on that server were encrypted as well. It didn't take long for the encryption to spread to each computer that was connected to the network. One by one, the dominos had started to fall. Files... email... phone system... ALL ENCRYPTED! No big deal, though—I have backups. Or so I thought.

"Ok, let's get started..." I told him. "Pull up your IRP and start assigning tasks."

"IRP?" he sheepishly replied.

"You know, an I-R-P... Incident Response Plan." I started to explain. "This is a structured set of procedures that guides an organization's response to cybersecurity incidents. It outlines steps to be taken before, during, and after a security breach to minimize impact on systems, data, and operations."

Based on the look on his face, I could tell that he did not have one in place. He started sweating right before my eyes. He had every reason to, right? I mean, he was the IT manager for this company, and they trusted him to keep their systems secure and up to date. The only documentation that I had for the company at the time was what I had compiled myself, and that was limited to setting up computers in Active Directory and loading the necessary programs. Luckily for him, I had been working on a template; an IT department "best practices" checklist that I would use to onboard new customers. There was a section on Incident Response Planning that would certainly prove useful to him in a time like this.

About a month prior to this incident, Tom had worked with a backup solution provider to upgrade the size of the current onsite backup appliance. This required sending the old one back once the new one was in place. Tom and I had held the old appliance for one week while we verified the backups were successfully going to the new appliance. Once we had completed a full backup and two incremental backups, we shipped the old appliance to the vendor. Tom received an email asking what he wanted to do with the data on the old appliance, with four options listed to choose from:

- Option 1: Migrate from cloud – Data and backup configuration are migrated from cloud.
- Option 2: Migrate from Old Device – Data and backup configuration are migrated from your old device.
- Option 3: Migrate Configuration Only – Your backup configuration is migrated, but data is not migrated.
- Option 4: Do not Migrate – Your data and backup configuration are not migrated

Since we had already configured the new appliance and verified successful backups, Tom replied to them with his choice—Option 4: Do not Migrate.

Well, this was apparently the wrong decision because the backup solution provider took that to mean that we did not want any of the data; none from the old device nor any from the device that had just been configured. So, when Tom went to the backup dashboard to start the restore process, he saw something that turned his face white as a ghost and made his stomach drop—no backups available.

If you have ever faced that type of scenario, you know how serious everything is. Not only are you dealing with the initial BEC, but now this company has no access to their data except what is on the users' laptops that were not on the network. One by one, the phone calls and text messages started cascading in from users who all wanted to know when they could get back in their email. The logistical fires were burning in every corner of the business, from Accounting and Payroll to Dispatch and Sales. Tom was in way over his head, so he reached out to me for assistance.

"Ok, let's get started..." I told him. "Pull up your IRP and start assigning tasks."

"IRP?" he sheepishly replied.

"You know, an I-R-P... Incident Response Plan," I started to explain. "This is a structured set of procedures that guides an organization's response to cybersecurity incidents. It outlines steps to be taken before, during, and after a security breach to minimize impact on systems, data, and operations."

Based on the look on his face, I could tell that he did not have one in place. He started sweating right before my eyes. He had every reason to, right? I mean, he was the IT manager for this company, and they trusted him to keep their systems secure and up to date. The only documentation that I had for the company at the time was what I had compiled myself, and that was limited to setting up computers in Active Directory and loading the necessary programs. Luckily for him, I had been working on a template; an IT department "best practices" checklist that I would use to onboard new customers. There was a section on Incident Response Planning that would certainly prove useful to him in a time like this.

An effective Incident Response Plan (IRP) can make all the

difference in how quickly your business can recover from a cyber-attack. When an attack happens, and trust me, they will, your team needs to know what tasks are assigned to them and when they should perform each task. Think back to grade school when you would have fire safety drills. The purpose of those drills was to make sure that everyone knew where to go and what to do in case of an actual fire. And, just like those drills prepared you for an actual fire, having a good IRP will prepare you for an actual attack.

The first item in our incident response plan would be a **Jump Bag**. A jump bag is a centralized document with critical information that incident responders will need, such as contact lists, escalation policies, technical documentation, and even on-call schedules, if applicable. Your contact list should include everyone in the company and any vendors that support your internal systems. The list should include the contact's first and last name, email address, phone number(s), and any relevant notes pertaining to their role in the IRP. Escalation policies should follow the chain of command in the organization, with department heads being the first point of escalation. From there the department head will notify the head of IT (or your MSP) about the incident, as well as anyone in direct leadership over them. Breaches are very serious because of the potential compromise of sensitive data, and as such, the Executive or C-level personnel should be included.

Technical documentation, or **Runbooks**, of your network infrastructure should outline the step-by-step procedures for returning the technical aspect of the business to normal working order. These will also help the teams respond faster and build a shared knowledge base.

What would IT be without a little chaos—am I right? You can't stop the chaos from happening, but you can prepare for it with what we call "**Chaos Engineering**." Just like those fire safety drills, you should experiment with system failures intentionally so your team can develop better responses and perhaps fault tolerance.

Most cybersecurity firms utilize a Network Operations Center, or NOC, to monitor traffic on the network. Using modern tools with AI to assist in spotting unauthorized access attempts, team members can quickly identify threats, isolate infected or compromised endpoints, and alert internal personnel of the incident in almost real-time responsiveness.

When it comes to alerts, you will want to implement some sort of aggregate software that will help to centralize alerts to filter noise and focus on critical matters. Make sure the alerts have as many details as possible to expedite the remediation process.

When a business is caught up in a breach or ransomware attack, the responses from your end users can be a bit overwhelming. You will want to establish **Incident Severity Levels** to prioritize the process. These severity levels (e.g., low, medium, high) allow the responders to determine which issue needs to be addressed first.

In the heat of the moment, when things are going so fast to make your head spin, communication and coordination are vital to making sure that everyone knows their responsibilities. In Tom's situation, the email server was also down, and that business lived and breathed out of their email. Cell phone communications were still functional, so Tom was able to send texts to all the employees to relay important information and instructions. Thankfully, he had already been using group texts in the service department, and it did not take us long to create the other groups.

Establishing clear communication channels for incident reporting will alleviate a little bit of that chaos. Before the incident, Tom had used SMS to communicate with just the service team. Since then, he has created SMS groups for each department so that he can communicate more effectively. It is important to define roles and responsibilities for team members during an event. Each department should have a designated person who is responsible for communications, typically the department head. That individual will relay instructions to their team as needed and inform the Incident Response Leader with status updates.

Privacy concerns are a politically popular topic for legislatures, and as such, those privacy concerns have led to the enactment of Breach Notification Obligations for all 50 states. Businesses are required to notify an individual of the breach if it is determined that certain types of information have been compromised. These are first name/initial and last name *plus* any of the following:

- Social Security Number (SSN)
- Driver's license number/state ID number
- Account number, credit or debit card number, in combination with any PIN, security code, access code, or password that would permit access to an individual's financial account.

If this type of information was breached, you should coordinate with legal counsel and the executive teams to draft your notice to your Secretary of State's office. In some cases, PR firms have been hired to help with the public disclosure.

Remember how Tom's company computers had mapped drives to the server? That is great for having a centralized data store that users can pull from, but it's also the fastest way to deliver the malware to the rest of the network. When a ransomware attack is detected, you must isolate the affected systems to prevent further damage. This can easily be handled by a firewall appliance that utilizes Zero Trust Network Access, or ZTNA. With ZTNA, you can define security policies for devices that connect to your network. If a device does not pass the security posture check performed by the firewall, then the failed device will instantly be isolated from the rest of the network.

After you have isolated the affected device, you will want to preserve as much evidence as you can for forensic analysis. Tom's company had purchased a cyber security insurance policy, and thankfully, they were covered for this kind of attack. The insurance company dispatched a team of their own to perform the forensic analysis. All Tom had to do was show them the affected devices and they took it from there.

While the insurance team was busy analyzing the systems, the rest of the employees still needed to be able to work. It's all great and wonderful that you have backups (bonus points if they are current), but it still takes a long time to restore data from a backup. All of that employee productivity would be wasted if you don't have a Business Continuity Plan (BCP). A BCP is a form of shadow IT where you have a virtual copy of the network environment ready to switch on in the event of an actual disaster. For some, this would be a virtual server environment resting on a non-production server, just receiving data from the production environment via data deduplication. Others have chosen a more modern method of adopting cloud services to ensure business continuity.

A Disaster Recovery Plan (DRP) is a set of instructions that help to return the business to its original operational state. This includes detailed documentation of the current technological inventory, including hardware and software assigned to each user. Tom's company did not have a standard computer for everyone; some computers had just enough resources to check email and surf the web, while others had 8-core CPUs. After the insurance team finished their analysis, computers were wiped, reloaded with the current version of Windows, and then configured for a user. Some users received a computer that was way more powerful than the one they had before, while others were struggling to run the simplest of programs. When creating a DRP, make sure that the computer is returned to the original user so as not to upset an already frustrated end-user. Soon after this attack, Tom implemented a standard computer package for each user in a specific department based on the computing needs of their job role.

The dust eventually settled, and the insurance investigators returned to their office to compile the report. Tom and I got right to work on implementing an IRP for this incident and collaborating on a few hypothetical doomsday scenarios. Before long, we had an extensive set of runbooks and IRPs for his business and many others. We had successfully navigated his company through the ransomware attack and returned the network to normal operating conditions. Security of the network going forward is critical, so it is best practice to have everyone change their passwords on a regular basis. If you are not using MFA (multi-factor authentication), then you should implement it now. Apps like Microsoft Authenticator or Google Authenticator are free and easy to configure.

Should you ever find yourself in Tom's shoes, just remember to take a deep breath and follow the IRP for that incident. Cyberattacks are not going to stop. If anything, they are only going to increase in severity. After every incident, you and your team should conduct a thorough review and identify what areas need improving. Don't think of your IRP as a static document—enshrined in glory and unchangeable. IRPs should be dynamic and adapt to emerging threats.

Steve Schulenberg
JCS Technologies, Inc.

STEVE SCHULENBERG has navigated a diverse career path, transitioning from sonar technician in the US Navy to founder of JCS Technologies. After his military service, Steve honed his skills in home construction and home entertainment technology before pivoting to IT consulting.

In 2014, he established JCS Technologies, a company dedicated to safeguarding small businesses in the digital age. A pivotal ransomware attack on a major client in 2017 shifted his focus entirely to cybersecurity. As a certified Microsoft Server Administrator and Engineer, Steve brings a wealth of expertise in network security, data protection, and cyber threat mitigation. He and his team specialize in implementing robust cybersecurity measures to protect businesses from evolving digital threats.

Steve lives in Indiana with his wife, Gretchen, and their son, Gavin.

To learn more or contact our team go to:
https://jcs-technologies.com/about-us/

SCAN THE QR CODE:

AFTERWORD
COMMON CYBERSECURITY MISCONCEPTIONS DEMYSTIFIED

Author: Ryan O'Hara
Company: 7 Figure MSP™

WHILE REGULATION IS COMING and will eventually require businesses to implement some form of cybersecurity in their business, no one should wait until they are forced to take action to protect themselves. Spinning up cybersecurity from nothing is not something you can do overnight. Taking steps now will not only ensure you are ready and prepared when regulation is enforced, but it will likely keep you from having to experience an attack in the meantime.

Here are some answers to the most frequent push-back responses from business owners:

- **"We're too small to be hacked."**

This is a common assumption by many small and medium-sized businesses. The reality is that you're just too small to make the news when it happens. Many smaller businesses also fail to report these attacks. Most statistics released in the past several years have shown that small and medium-sized businesses are not only hacked frequently, but they also make up the majority of reported cyberattacks. Hackers rarely target their victims. They usually employ strategies like net fishing. They cast a wide net and go after the low-hanging fruit. Small and medium-sized businesses typically have little to no real cybersecurity controls and are ripe for the picking.

- **"I don't know anyone who's been hacked."**

You almost certainly do. Business owners tend to know and interact with a lot of other business owners. The reason that you "don't know any other business owners who have been hacked" is that they don't talk about it. Getting hacked is not a fun experience, and even minor attacks come with a stigma attached. Usually, these attacks start with someone clicking on something or otherwise doing something that, in hindsight, was a bad idea. People just don't talk about these things out of embarrassment.

- **"I don't have anything the hackers want."**

First, you probably do. There are many ways hackers can profit financially from an attack. Ransomware can cripple your computers to the point that you can't operate your business and may pay the attackers just to regain access to your machines and data. Business Email Compromise can result in an interception of financial information intended for ACH or wire transactions diverting funds meant for you or a vendor to the hacker's account. Even if you don't deal with customers' private information, you do have social security numbers for your employees that you use for payroll.

Second, since hackers rarely target specific businesses and rely more on net fishing-type attacks, they don't even know who you are until they gain access and can poke around. Even if they find a way of getting $5,000 out of a small business, it's worth it. Hitting 100 businesses with little to no protection for small amounts is still more lucrative than the expense it takes them to get through the security of the bigger guys.

- **"I'm in the cloud, so I'm safe."**

Contrary to popular belief, most cloud providers have exclusions in their terms of service for situations where your data is lost or compromised. Most of them even recommend having third-party monitoring and backup of the service as they are not responsible. Additionally, most attacks that compromise data in the cloud are not compromises of the cloud provider themselves. Often, the compromise is due to a user's credentials being obtained by the attacker and being used to log in to the platform legitimately. If the hackers used your username and password to access the data, it's not the provider's fault, regardless of what their terms and conditions say.

Another issue with this assumption is that while most cloud providers have security controls available on their platform, they are often not enabled by default. This means it's your responsibility to turn them on and use them and without the guidance of a qualified IT or Cybersecurity professional, most don't know they are there, much less enable them.

- **"My insurance will protect me."**

Insurance is not in the business of protection. Insurance is there as a safety net to help pay for the costs of remediation in the event of a breach. Much like failing to wear a seatbelt in an auto accident, if you aren't doing enough to prevent an attack in the first place, your insurance will likely deny your claim entirely. Most insurance carriers require 10+ page questionnaires about the protections you have implemented before even offering cybersecurity policies to businesses. The fact is, to have your cybersecurity insurance policy payout, you need to have been doing enough to reasonably protect yourself in the first place.

- **"I'll just use paper and pencil."**

While in many businesses this may be possible in the short term, it's just not possible in the long term. Even if you can run your business without technology, it's highly unlikely that you can interact with all the vendors required for you to operate without it. If you experience a cyberattack and can continue to operate temporarily via paper and pencil, someone is eventually going to have to enter that information once your computers are back up. So, while you may not experience "down time," you will still have a labor loss due to the additional time someone will have to get all that paper/pencil information back into your systems.

- **"I'll just close my business."**

Sometimes when faced with the reality of the added cost required for proper cybersecurity, business owners just say, "If I get hacked, I'll just close my doors." This is a gut reaction to a reality that they likely hadn't considered until that moment. Yes, you can close your business, but you likely depend on that business for your livelihood, that of your family, and your employees as well. Is it worth rolling the dice? Another consideration is the legal ramifications of a breach. Closing your business may not protect you from lawsuits brought on you for failure to protect customers' information.

The moral of the story is you never want to have to go through the experience of a cyberattack. Even in the best of circumstances, where your business can recover, and your insurance company pays out on your claim, it is not a pleasurable experience. The ramifications of a cyberattack also take months or years to fully recover from.

No one can promise 100% protection, but implementing proper cybersecurity can significantly reduce the likelihood of an attack. It will also likely significantly reduce the damage caused by the attack and the subsequent recovery time. Finally, having adequate protections in place before an attack will allow you to make a claim with your cybersecurity insurance provider that will most likely get paid out.

Thank you for reading *Technology Legislation is Coming.*

Made in United States
North Haven, CT
10 November 2026

22084529R00088